ADVANCES IN NUTRITION AND CANCER

ADVANCES IN EXPERIMENTAL MEDICINE AND BIOLOGY

Recent Volumes in this Series

Volume 342
CORONAVIRUSES: Molecular Biology and Virus–Host Interactions
Edited by Hubert Laude and Jean-François Vautherot

Volume 343
CURRENT DIRECTIONS IN INSULIN-LIKE GROWTH FACTOR RESEARCH
Edited by Derek LeRoith and Mohan K. Raizada

Volume 344
MECHANISMS OF PLATELET ACTIVATION AND CONTROL
Edited by Kalwant S. Authi, Steve P. Watson, and Vijay V. Kakkar

Volume 345
OXYGEN TRANSPORT TO TISSUE XV
Edited by Peter Vaupel, Rolf Zander, and Duane F. Bruley

Volume 346
INTERACTIVE PHENOMENA IN THE CARDIAC SYSTEM
Edited by Samuel Sideman and Rafael Beyar

Volume 347
IMMUNOBIOLOGY OF PROTEINS AND PEPTIDES VII: Unwanted Immune Responses
Edited by M. Zouhair Atassi

Volume 348
ADVANCES IN NUTRITION AND CANCER
Edited by Vincenzo Zappia, Marco Salvatore, and Fulvio Della Ragione

Volume 349
ANTIMICROBIAL SUSCEPTIBILITY TESTING: Critical Issues for the 90s
Edited by James A. Poupard, Lori R. Walsh, and Bruce Kleger

Volume 350
LACRIMAL GLAND, TEAR FILM, AND DRY EYE SYNDROMES: Basic Science and Clinical Relevance
Edited by David A. Sullivan

ADVANCES IN NUTRITION AND CANCER

Edited by

Vincenzo Zappia

Institute of Biochemistry of Macromolecules
Medical School
Second University of Naples
Naples, Italy

Marco Salvatore

National Cancer Institute
Fondazione "Giovanni Pascale"
Naples, Italy

and

Fulvio Della Ragione

Institute of Biochemistry of Macromolecules
Medical School
Second University of Naples
Naples, Italy

SPRINGER SCIENCE+BUSINESS MEDIA, LLC

Library of Congress Cataloging-in-Publication Data

Advances in nutrition and cancer / edited by Vincenzo Zappia, Marco Salvatore, and Fulvio Della Ragione.
p. cm. -- (Advances in experimental medicine and biology ; v. 348)
Includes bibliographical references and index.
ISBN 978-1-4613-6278-4 ISBN 978-1-4615-2942-2 (eBook)
DOI 10.1007/978-1-4615-2942-2
1. Cancer--Nutritional aspects--Congresses. I. Zappia, Vincenzio. II. Salvatore, Marco. III. Della Ragione, Fulvio. IV. International Conference on Nutrition and Cancer (1992 : Naples, Italy) V. Series.
[DNLM: 1. Neoplasms--etiology--congresses. 2. Neoplasms--epidemiology--congresses. 3. Cell Transformation, Neoplastic--genetics--congresses. 4. Diet--congresses. 5. Risk Factors--congresses. W1 AD559 v. 348 1993 / QZ 202 A2445 1992]
RC268.45.A38 1993
616.99'4071--dc20
DNLM/DLC
for Library of Congress 93-42147
CIP

Proceedings of an International Conference on Nutrition and Cancer, held November 20–21, 1992, in Naples, Italy

ISBN 978-1-4613-6278-4

PREFACE

This volume contains the scientific contributions presented at the International Symposium held in Naples, Italy, in November 1992 at the National Tumor Institute "Fondazione Pascale". The Meeting gathered together experts from different disciplines, all involved in the vital and timely subject of Nutrition and Cancer.

About 15 years ago a consensus among cancer epidemiologists began to emerge suggesting that diet might be responsible for 30-60% of the cancers in the developed world. The best estimate, reported in a now classical paper by Richard Doll and Richard Peto (1981), was that by dietary modification, it would be possible to reduce fatal cancers by about 35%. Within about six years there was widespread agreement that the principal changes required were a reduction in consumption of fat, along with an increase in the consumption of fruit, green and yellow vegetables, dietary fiber, and some micronutrients. Attention was also paid to the methods of cooking and preservation of foodstuffs. On the other hand very few, if any, effects were attributed to food additives and to pollution of food by trace pesticides, to which the general public often gives unfounded importance.

All these aspects of the relationship between cancer and nutrition were a matter of discussion during the meeting and the relative contributions represent the main body of this book. In particular both epidemiological studies on the dietary risk factors (alcohol, fat, etc.) and the clinical aspects of main cancers due to mutagens occurring in the diet (gastric, colon, and breast tumors) have been analyzed in detail. Space has also been given to the now well-recognized role of the Mediterranean diet in the prevention of cancer.

The controversy on the relationship between cancer and dietary fibers was the subject of a heated round table discussion, reported here in one chapter. Indeed, failure to document a consistent inverse relationship with dietary fiber has led to research into the possible role of individual constituents of fibers. What promises to be more fruitful, is the recognition that some starch is also resistant to digestion.

A separate section of the volume concerns the central problem of the molecular bases of human cancerogenesis. It is now well accepted that malignant transformation is due to mutations that modify the mechanisms regulating normal cellular growth and development. These alterations include the somatic activation of cancer-promoting genes (cellular oncogenes) and the germline or somatic inactivation of tumor suppressor genes, also known as antioncogenes or recessive oncogenes. Much is now known in this field, in large part due to the enormous developments in molecular genetic techniques. It is notable in this context that the former separation of the three areas of basic cancer research, namely that related to chemical and radiation cancerogenesis, that of oncologic virology, and that of cytogenetic studies, has now been totally surpassed. The emergence of a joint, inter-disciplinary approach has already yielded a rich harvest of basic knowledge concerning cancer development, and will provide the seeds for future breakthroughs in diagnostic and clinical progress.

Vincenzo Zappia

ACKNOWLEDGEMENTS

The International Symposium on Nutrition and Cancer was held under the auspices of the Consiglio Nazionale delle Ricerche (Comitato Biotecnologie), the Società Italiana di Biochimica, and the Centro Internazionale di Studi sull'Alimentazione (Faenza), and was sponsored by the Istituto di Scienze dell'Alimentazione di Avellino del CNR, by Beckman Analytical S.p.A., and by Pierrel S.p.A.

Particular thanks are due to Rosanna Palumbo and Rosa Ruggiero of the Organizing Committee at the Institute of Biochemistry of Macromolecules of the University of Naples, and to Armando Tripodi and his staff at the "Pascale" Tumor Institute, where the Meeting was held.

The Editors would like to express thair gratitude to the authors of the articles and to Plenum Press for having made possible the publication of this volume.

Patricia Reynolds acted as Editorial Assistant, single-handedly guiding the project through all its phases, to render this "camera-ready" publication a compact and homogeneous volume.

CONTENTS

CLINICAL RESEARCH AND PERSPECTIVES

MOLECULAR BASES OF MALIGNANT TRANSFORMATION

GROWTH FACTORS AND MALIGNANT TRANSFORMATION

Stuart A. Aaronson, Toru Miki, Kimberly Meyers, and Andrew Chan

Laboratory of Cellular and Molecular Biology
National Cancer Institute
Bethesda, MD 20892 USA

INTRODUCTION

In the early 1980's, approaches aimed at identifying the functions of retroviral oncogenes converged with efforts to investigate normal mitogenic signaling by growth factors. A number of retroviral oncogene products were found to be similar to the protein kinase encoded by v-*src* product[1]. Unlike many protein kinases that phosphorylate serine or threonine residues, the v-*src* product is a protein kinase that specifically phosphorylates tyrosine residues[2]. Purification and sequencing of growth factors and their receptors revealed that the platelet derived growth factor (PDGF) B-chain is similar to the predicted v-*sis* oncogene product[3] and that the v-*erb*B oncogene product, which has sequence similarity to the v-*src* product, is a truncated form of the EGF receptor[4]. Binding of EGF to its receptor results in autophosphorylation of the receptor on tyrosine[5]. Oncogenes activated by a variety of mechanisms[6] frequently have been shown to encode growth factors, receptor tyrosine kinases or downstream effectors.

Growth Factor Requirements for Cell Proliferation

Growth factors cause cells in the resting or G_0 phase to enter and proceed through the cell cycle. The mitogenic response occurs in two parts; the quiescent cell must first be advanced into the G_1 phase of the cell cycle by "competence" factors, traverse the G_1 phase, and they become committed to DNA synthesis under the influence of "progression" factors[7]. Transition through the G_1 phase requires sustained growth factor stimulation over a period of several hours (Fig. 1). If the signal is disrupted for a short period of time, the cell reverts to the G_0 state[8]. There is also a critical period in G_1 during which simultaneous stimulation by both factors is needed to allow progression through the cell cycle[9,10]. After this restriction point, only the presence of a "progression" factor such as insulin-like growth factor 1 (IGF-1) is needed[11]. Cytokines such as transforming growth factor β (TGF β), interferon, or tumor necrosis factor (TNF) can antagonize the proliferative effects of growth factors. In the case of TGFβ, these effects can be observed even when added relatively late in G_1[12].

Advances in Nutrition and Cancer, Edited by
V. Zappia *et al.*, Plenum Press, New York, 1993

In some cell types, the absence of growth factor stimulation causes the rapid onset of programmed cell death or apoptosis[13]. Certain growth factors can also promote differentiation of a progenitor cell, while at the same time stimulating proliferation; others acting on the same cell induce only proliferation[14]. Thus, there must be specific biochemical signals responsible for differentiation that only certain factors can trigger[15,16]. The actions of a sequential series of growth factors can cause a hematopoietic progenitor to move through stages to a terminally differentiated phenotype[14]. However, at intermediate stages, in the absence of continued stimulation by the factor, this commitment is not irreversible[17]. Although the differentiation program of the cell governs the diversity of phenotypic responses elicited, there are some common highly conserved downstream effectors of mitogenic signaling. For example, introduction of foreign receptors by DNA transfection into cells often allows coupling of the appropriate ligand to mitogenic signal transduction pathways inherently expressed by the cells[18].

Receptor Tyrosine Kinases and their Effectors

Cells of most if not all major tissue types are targets of growth factors that mediate their effects by means of receptors with intrinsic tyrosine kinase activity. These receptors have an extracellular ligand binding domain and an intracellular tyrosine kinase domain

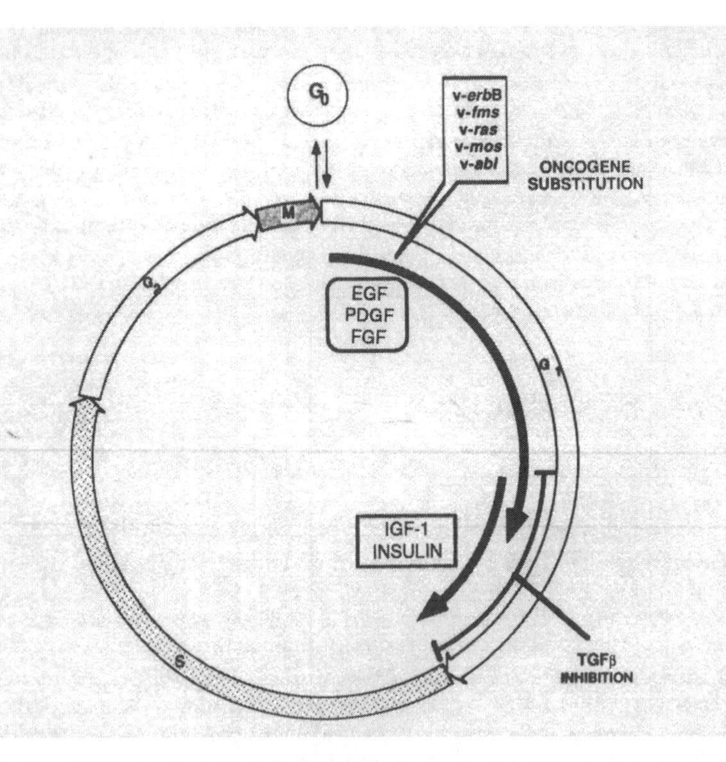

Figure 1. Growth factor requirements during the cell cycle. A schematic representation of requrements for the coordinated actions of two complementing growth factors to induce cell DNA synthesis. In BALB-MK cells, several oncogenes can specifically substitute for the competence factor requirements. The ability substitute for the competence factor requirement. The ability of TGFβ to inhibit the onset of DNA synthesis, even when added in late G_1 is also depicted.

responsible for transducing the mitogenic signal (Fig. 2). Ligand binding induces formation of receptor dimers or oligomers[19]; molecular interactions between adjacent cytoplasmic domains lead to activation of kinase function. Most evidence indicates that the transmembrane domain does not directly influence signal transduction and is instead a passive anchor of the receptor to the membrane. However, point mutations in the transmembrane domain of one receptor-like protein, the *neu/erb*B-2 protein, enhances its transforming properties[20]. The tyrosine kinase domain is the most conserved and is absolutely required for receptor signaling. Mutation of a single lysine in the ATP binding site, which destroys the ability of the receptor to phosphorylate tyrosine residues, completely inactivates biological function[21]. The receptor itself is often the major tyrosine

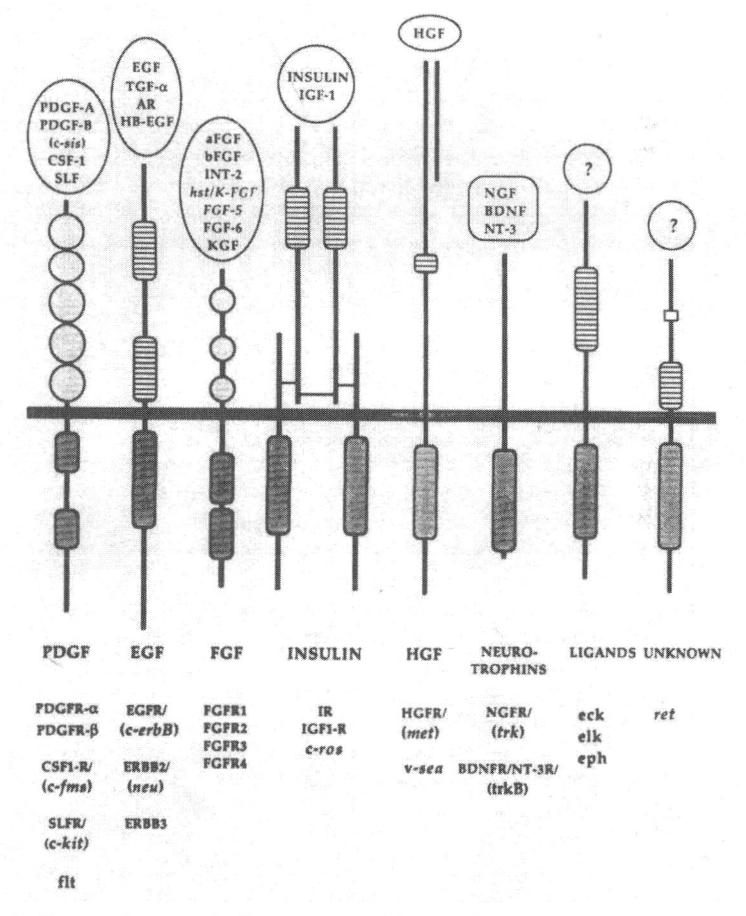

Figure 2. Transmembrane tyrosine kinases. Structural features of various receptor tyrosine kinase receptors are shown. Each receptor family is designated by a prototype ligand. Growth factors known to bind to receptors of a given family are listed above, and receptors that constitute each family are listed below. Boxes denote those growth factors or receptors whose genes were initially identified as activated oncogenes. The c-*onc* designation is used to specify cellular homologs of retroviral oncogenes. Open circles illustrate immunoglobulin-like repeats. Dashed boxes indicate cysteine-rich domains. Dotted boxes indicate conserved tyrosine kinase domains.

phosphorylated species observed following ligand stimulation. Tyrosine phosphorylation may modulate kinase activity, but certainly affects the ability of the kinase to interact with substrates[22]. Figure 2 illustrates several distinct families containing structurally related members. A number are encoded by protooncogenes, whose viral counterparts, like v-*sis* and v-*erb*B, were initially identified as retroviral oncogenes[23], or activated by retroviral integration[24]. Others were detected as cellular oncogenes by DNA transfection[25]. Still others reflect genes molecularly cloned on the basis of structural similarity to other tyrosine kinases[26,27] or by identification of protein sequence[28]. The spacing of cysteine residues in their external domains defines either immunoglobulin-like domains in the case of PDGF and FGF receptor families, or cysteine-rich clusters. The known ligands for each receptor family also show similarities in cysteine spacing despite otherwise considerable divergence[29]. Recently identified ligand-receptor systems include hepatocyte growth factor (HGF)-*met* [30] and NGF-*trk* [31] as well as two NGF related ligands, brain derived neurotrophic factor and neurotrophin-3, which interact with the *trk*-B product[32].

The PDGF system has served as the prototype for identification of substrates of the receptor tyrosine kinases. Certain enzymes become physically associated and are phosphorylated by the activated PDGF receptor kinase. These proteins include phospholipase C (PLC-γ)[33], phosphatidylinositol 3' kinase (PI-3K)[34], *ras* guanosine triphosphatase (GTPase) activating protein (GAP)[35], *src* and *src*-like tyrosine kinases[36]. These molecules contain noncatalytic domains called *src* homology (SH) regions 2 and 3. SH2 domains bind preferentially to tyrosine phosphorylated proteins and SH3 domain may promote binding to membranes or the cytoskeleton[22]. The *raf* proto-oncogene product has also been reported to become physically associated with the receptor and tyrosine phosphorylated as well[37] although it lacks SH2 or SH3 domains (Fig. 3).

PLC-γ is one of several PLC isoforms. Its hydrolysis of phosphatidylinositol [4,5] bis-phosphatase generates two second messengers, inositol triphosphate and diacylglycerol[38]. The former causes release of stored intracellular calcium and the latter activates protein kinase C (PKC). These second messengers appear rapidly in cells after stimulation by growth factors such as PDGF. The relative increase in their synthesis *in vivo* correlates with the ability of a particular receptor kinase to induce tyrosine phosphorylation of PLC-γ[33]. Phosphorylation of PLC-γ on tyrosine increases its catalytic activity *in vitro*[39]. Together, the results indicate that receptor induced tyrosine phosphorylation activates PLC-γ. The actions of a number of tumor promoters are thought to be mediated by PKC[38].

PI-3K, which phosphorylates the inositol ring of phosphatidylinositol in the 3' position, becomes physically associated with a number of activated tyrosine kinases[40]. The 85-kDsubunit of the protein contains two SH2 domains and an SH3 domain and is tyrosine phosphorylated, but the subunit lacks PI-3K activity[41]. The catalytic domain is likely associated with a 110-kD protein that is part of a heterodimeric complex with the 85-kD protein[40]. The transforming ability of polyoma middle T mutants correlates with the functional activity of PI-3K in complexes with pp60$^{c\text{-}src}$ [42]. Moreover, v-*src* and v-*abl* mutants that fail to associate with PI-3K are non-transforming[43]. Thus, PI-3K may function in the process of transformation.

GAP is intimately involved in the function of the *ras* proteins[44]. It stimulates the GTPase activity of the *ras* gene, a 21-kD guanine nucleotide binding protein (Ras)[45]. Ras is a critical component of intracellular mitogenic signaling pathways. Microinjection of oncogenically activated Ras into NIH/3T3 fibroblasts induces DNA synthesis[46]. GAP acts as a negative regulator of *ras* function[47]. Mutations that cause oncogenic activation of Ras lead to accumulation of Ras-bound to GTP, the active form of the molecule[45]. These mutations in *ras* block the ability of GAP to downregulate Ras to its inactive GDP-bound form[44]. GAP may also function in a complex with Ras as an effector of its downstream signaling functions[49]. Thus, mutations which impair interaction of Ras with GAP also block the biological function of Ras.

Stimulation of certain receptors results in physical interaction of GAP with the receptor kinase[35] as well as its association with the cell membrane, the known site of Ras function[45]. Tyrosine-phosphorylated GAP is also found to be associated in a complex with at least two other tyrosine phosphorylated proteins (p62 and p190) that may modulate Ras function[49]. Interaction with p190 decreases the ability of GAP to promote GTPase activity of Ras *in vitro*[50]. Stimulation of cells with PDGF leads to an increase in the amount of GTP-bound Ras[51], consistent with the possibility that tyrosine phosphorylation of GAP associated proteins transiently interrupts its inhibition of Ras function. However, other proteins that regulate Ras including a protein that promotes release of bound GDP in exchange for GTP have been identified in yeast[52], and there is evidence for such activities in mammalian cells as well[53]. Thus, further studies are needed to firmly establish the mechanisms responsible for activation of Ras in growth factor-stimulated cells as well as the effector functions of this important signal transducer.

The *src* gene and structurally related family members including *yes* and *fgr* were initially identified as oncogenes of retroviruses. These and the other members of the *src* family encode nonmembrane spanning tyrosine kinases[54]. Some show expression only in certain differentiated cell types, consistent with their possessing highly specialized functions, while others are widely expressed. *Src* and other related tyrosine kinases are activated rapidly in cells stimulated with PDGF[36].

The *raf* proto-oncogene product is a serine-threonine kinase[55] that is activated by a PKC-independent mechanism in response to a number of growth factors[56]. Oncogenically activated forms of Raf resulting from deletions or mutations of its NH_2-terminal domain have been identified in tumors by means of gene transfer experiments. The NH_2 terminal domain may normally serve to regulate the catalytic domain. Thus, *raf* oncogenes show constitutively increased serine-threonine kinase activity. The substrates of *raf* remain to be identified.

Oncogene Subversion of Specific Signaling Pathways

The evidence summarized above indicates that proto-oncogene products act at critical steps in growth factor signaling pathways. Thus, their constitutive activation as oncogene products would be expected to profoundly influence cell proliferation and possibly the differentiated state of the transformed cell. Tumor cells exhibit reduced requirements for serum in culture. The actions of oncogenes have been investigated with respect to their ability to subvert the actions of the two major growth factor signaling cascades. For instance, mouse keratinocytes can be propagated in chemically defined medium containing only two complementing growth factors, EGF and IGF-1[57]. Introduction of various oncogenically activated receptor kinases or *ras* or *raf* oncogenes completely alleviate the requirement for EGF but not IGF-1 (Fig. 1). These findings support the concept that the signaling pathway of competence factors ordinarily limits growth *in vivo* and that genetic changes activating critical regulatory molecules within this pathway are commonly selected during evolution of the malignant cell.

The state of differentiation of a cell can influence the action of an oncogene and affect the phenotype of the transformed cell. For example, the time of expression of the *ras* oncogene during differentiation of keratinocytes determines whether or not malignancy is induced[58]. PC12 neural cells undergo terminal differentiation in response to *ras* or *src* oncogenes[59], and introduction of *ras* into EBV-immortalized B lymphoid cells induces plasma cell differentiation[60].

Implication of Other Mitogenic Signaling Systems in Malignancy

Binding of ligands to at least two classes of receptors distinct from membrane spanning tyrosine kinases is known to stimulate cell proliferation. One class includes the receptor for Interleukin-2 (IL-2), IL-3, IL-4, IL-6, IL-7, granulocyte-macrophage colony-stimulating factor (GM-CSF), G-CSF and erythropoietin (Epo). These receptors are membrane glycoproteins with a single hydrophobic transmembrane domain[61]. Their external domains are similar in size and contain several conserved cysteines in their NH_2-terminal portions. In contrast, their cytoplasmic domains vary in length, show little if any sequence similarity, and possess no tyrosine kinase domain. Some of these receptors require associated proteins for high affinity ligand binding[62].

Little is known of the biochemical pathways by which these receptors stimulate proliferation, although their activation can lead to the appearance of tyrosine phosphorylated proteins[63] and increased amounts of GTP-bound Ras[64]. Binding of IL-2 to its receptor activates Lck[65]. Thus, the Src family of tyrosine kinases (which includes Lck) may participate in signal transduction by this class of receptors. Certain *in vitro* mutations of the Epo receptor constitutively activate the receptor and cause transformation of appropriate hematopoietic target cells[66]. Erythroblastic leukemia induced by the spleen focus-forming virus is due to molecular mimicry of Epo by a recombinant *env* gene product of this defective retrovirus[67]. In human T cell tumors associated with HTLV-1 infection, viral gene products appear to stimulate proliferation of affected cells by increasing expression of both IL-2 and its receptor[68].

Another class of molecules capable of causing mitogenic stimulation of certain cell types are neurotransmitters. The topography of these receptors includes, in addition to their seven transmembrane domains, an extracellular NH_2-terminal domain, and a cytoplasmic COOH-terminal tail or large intracellular loop containing regulatory serine-threonine residues. The heterodimeric guanine nucleotide binding proteins (G proteins) activated by such receptors can be coupled to various effectors including adenylyl cyclase, phospholipase C, and K_+ channels[69]. These receptors may also stimulate tyrosine phosphorylation, but the kinases responsible remain to be identified[70]. Related genes encode the α_1, α_2-, β_1-, and β_2-adrenergic receptors, the muscarinic acetylcholine receptors (mACHR), the serotonin receptors, the substance K receptor, the dopamine receptors, the bombesin receptor and the endothelin receptor[71]. The *mas* oncogene isolated by gene transfer from a human carcinoma encodes a seven membrane spanning receptor[72]. Overexpression of certain acetylcholine or serotonin receptor subtypes after transfection in NIH/3T3 cells, causes ligand dependent transformation[73].

Bombesin-like peptides are secreted by neuroectodermally derived small cell lung carcinomas and stimulate growth of these cells[74]. Moreover, antibodies to bombesin have been reported to inhibit tumor cell proliferation *in vitro* and *in vivo*[75]. These findings raise the possibility that autocrine stimulation by ligands for other G protein coupled receptors may occur in tumors as well. It would follow that genes that act at rate limiting steps in signal transduction might be subject to oncogenic activation in specialized cell types in which neurotransmitters are normally mitogenic. Indeed, this appears to be the case. For instance, growth hormone-secreting pituitary tumors and endocrine tumors of the adrenal cortex and ovary frequently exhibit point mutations in G proteins that interact with adenylyl cyclase[76]. The affected residues, analogous to those activating *ras*, would lead to constitutive activity and increased intracellular concentrations of cAMP.

Expression cDNA Cloning of Growth Regulatory Genes

The ability to identify human oncogenes has to a large extent been limited by assay techniques. For instance, mutated forms of hematopoietic growth factor receptors efficiently couple with mitogenic signaling pathways and induce transformation in a hematopoietic progenitor cell, but fail to do so in NIH/3T3 fibroblasts frequently used for detection of transforming genes by transfection[66]. Similarly, overexpression of only G protein coupled receptors linked to phosphtidylinositol turnover, but not those coupled to adenylyl cyclase, appears to cause ligand dependent transformation of NIH/3T3 cells[77] To detect an oncogene by gene transfer, it must be small enough to be transfected and its promoter must allow a high level of expression in the recipient cell. Some of these problems have recently been overcome by the development of efficient cloning vectors allowing stable expression[78]

Expression Cloning of a Transforming Gene from a Human Sarcoma cDNA Library

An expression cDNA library was constructed from a pool of poly(A)$^+$ RNA derived from two human tumor cell lines, A2095 and RD-ES-1. Following transfection of the library DNA into NIH/3T3 cells, four morphologically distinct foci were selected for further analysis. To identify the cDNA sequences responsible for the transformed phenotypes, integrated plasmid DNAs were first released from the mouse genome by *NotI* digestion followed by circularization for transformation of bacterial cells. Individual rescued plasmids were then tested for their transforming potential in NIH/3T3 cells. This strategy led to the isolation of a plasmid, 58-1, that demonstrated high-titered transforming activity (>10^4 focus-forming units/pmol) on NIH/3T3 cells.

Clone 58-1 contained a ~4.0 kb cDNA insert. Sequence analysis from its 5' end revealed strong similarity to a recently cloned mouse cDNA which encodes for an a subunit of a G protein, Gα12^{mu}[79]. Available information derived from the mouse sequence indicated that clone 58-1 possessed the entire coding region in addition to 6 bp of 5' untranslated region and ~3 kb of 3' untranslated sequence The predicted amino acid sequence of clone 58-1 diverged from Gα12^{mu} in only six residues, strongly suggesting that it was the human homolog (Gα12^{hu}) of the mouse sequence[80] To identify the translational product of this cDNA clone, we performed *in vitro* transcription-translation analysis. This study revealed a major protein species with a relative molecular mass of ~45 kDa, consistent with the size of 44 kDa calcuated from its predicted coding sequence and resembling the sizes of other a subunits of known high-molecular weight G proteins

Transforming properties of Gα12

To ascertain whether Gα12^{hu} cDNA could act as a classical oncogene, plasmid 58-1 was transfected into NIH/3T3 cells, and marker-selected mass cultures were obtained. As controls, parallel cultures were transfected with a known oncogene, c-*sis*/PDGF-B, and with the vector alone. Gα12^{hu} transformed foci were stellate in appearance, with aggregation of cells at high density in the center of the focus, resulting in an overall punctuate morphology. Gα12^{hu} transformants also showed a decreased doubling time and increased saturation density compared with control pSV2neo transfectants (Table 1). Another property of transformed cells is the ability to grow in semisolid medium. As shown in Table 1, Gα12^{hu}-transformed cells formed large, progressively growing colonies in soft agar. Finally, inoculation of Gα12^{hu} transfectants subcutaneously into athymic nude mice induced tumor formation at high frequency, while transfectants containing the control plasmid pSV2neo had substantially lower incidence of tumors under the same assay conditions (Table 1). All of these findings indicated that $G\alpha12^{hu}$ transformation of NIH/3T3 cells both induced morphological alterations and enhanced proliferation both *in vitro* and *in vivo*.

Overexpression of Wild-Type Gα12 is Sufficient for Transformation

Because our $G\alpha12^{hu}$ cDNA clone was derived from human tumor cells, we investigated whether its oncogenic activation was due to mutations within the coding sequence as has been reported for several other a subunits of G proteins[76]. On the basis of our sequence, we identified six amino acid positions that showed divergence between the human and mouse sequences. Of these, three represented conservative changes (serine to threonine, lysine to arginine, and isoleucine to valine). Others included substitution of histidine 119 by tyrosine and of alanine 26 by serine and glycine as well as a glycine addition in position 18 (Fig. 1). To determine whether these disparities were due to species differences or tumor-specific mutations, we screened a cDNA library constructed from a normal human mammary epithelial cell line[81] for additional $G\alpha12^{hu}$ cDNAs. A total of 33 positive clones were isolated, and cDNAs that contained the entire coding region were identified by polymerase chain reaction. Sequence analysis revealed no differences at the candidate sites for mutations between cDNA clones derived from either normal or tumor

Table 1. Transforming properties of NIH/3T3 transfectants.

Plasmid	Transforming frequency[a] (FFU/pmol)	Soft agar[b] growth (%)	Cell dou-[c] bling time (h)	Saturation[d] density	Tumorigenicity[e] (frequency)
pSV2neo	<1.0	1.2	30	2.3	1/6
sis	18×10^4	21.2	26	4.5	5/5
$G\alpha12^{hu}$	9.8×10^4	19.9	23	4.8	6/6

[a]NIH/3T3 cells were transfected with ~0.01 mg of each plasmid, and the number of focus-forming units (FFU) was scored after 3 weeks in cultures. All three plasmids produced similar numbers of marker-selectable colonies.

[b]NIH/3T3 transfectants were suspended in 0.4% soft top agar in the presence of 10% CS. Colonies of >0.2 mm were scored after 14 days, and data represent average values of duplicate plates.

[c]Approximately 3×10^4 cells were plated in duplicate in 60-mm-diameter plates and cultured in the presence of 5% CS. Cells were counted each day for 6 consecutive days, and the result was used in the calculation of exponential doubling time.

[d]NIH/3T3 transfectants were plated as described in footnote [c]. Cells were allowed to grow to confluence until cell numbers did not alter after three consecutive counts. Data represent average values of duplicate samples.

[e]Approximately 1×10^5 to 2.5×10^5 cells were introduced subcutaneously into athymic nude mice. Data indicate incidence of tumors 5 weeks after inoculation. All tumors generated by $G\alpha12^{hu}$ were of well-differentiated fibrosarcomas.

cell lines. Moreover, examination of the biological activities of several normal cDNA clones revealed that each exhibited high-titered transforming activity comparable to that of clone 58-1 (data not shown).

DISCUSSION

The prevalent view of human carcinogenesis postulates a multistep process involving the activation of cellular proto-oncogenes and inactivation of tumor suppressor genes[82,83]. By means of genomic DNA transfection-transformation assays utilizing NIH/3T3 mouse fibroblasts, various oncogenes have been isolated from both human and rodent tumors. They represent diverse classes of growth-regulatory molecules, including growth factors, growth factor receptors, mitogenic signal transducers, and transcription factors[6]. Frequently, mutational alterations inflicted on coding sequences or transcriptional elements of normal cellular proto-oncogenes have been shown to be responsible for oncogenic activation. This approach has generated substantial knowledge concerning the molecular mechanisms of malignant transformation but is limited with respect to oncogene detection. First, oncogenic sequences of considerable size (>100 kb) may not be detected because of inefficiencies in gene transfer techniques[84]. Second, promoter/enhancer elements of transforming genes may not function optimally in fibroblast cells. To overcome these inherent difficulties, we have taken advantage of an expression cloning system in which a high proportion of full-length cDNA can be synthesized and cloned directionally into a phagemid expression cloning vector containing a strong retroviral promoter[78]. Following transfection of library cDNA into NIH/3T3 cells, transformed foci can be identified, and plasmids containing oncogenic sequences can be efficiently rescued by using this vector system[78].

We describe the isolation of transforming cDNAs from a soft tissue sarcoma-derived expression library. This cDNA encoded a member of a new class of $G\alpha$ subunits designated $G\alpha12$. Our findings demonstrate that at least one member of this class of $G\alpha$ has the ability to act as an oncogene in inducing transformed foci in rodent fibroblasts, as well as anchorage-independent growth and tumor formation in animals. The possibility that mutational activation was responsible for its transforming activity was excluded by our demonstration of the same sequence of the $G\alpha12^{hu}$ coding region in cDNAs isolated from a normal human epithelial cell library as well as the highly efficient transforming activity of $G\alpha12^{hu}$ cDNAs derived from both tumor and normal cell libraries.

We have no direct evidence that $G\alpha12^{hu}$ overexpression was involved in the malignant conversion of either of the soft tissue sarcomas from which the cDNA library was generated. One sarcoma, RD-ES-1, showed a higher level of $G\alpha12^{hu}$ transcript than did any of the normal cell types analyzed, but this level was significantly lower than the level expressed by $G\alpha12^{hu}$-transformed NIH/3T3 cells. If this gene was overexpressed in RD-ES-1 cells, we detected no evidence of any gross gene rearrangement by Southern blot analysis[85] that might account for increased transcript levels. Nonetheless, our present findings demonstrating the oncogenic potential of this new class of Ga subunits warrants a search for evidence of genetic alterations that cause overexpression or constitutive activation of this gene in human malignancies.

The stable expression cloning approaches summarized here have proven effective in detecting putative oncogenes that act at limiting steps in mitogenic signaling pathways. It seems likely that as these approaches are combined with efforts to increase the efficiency of stable transfection of a variety of recipient cells, the number of molecules identified as being critically involved in growth factor signaling and cancer will expand.

REFERENCES

1. M. Collert and R. Erickson, Protein kinase activity associated with the avian sarcoma virus *src* gene product, *Proc. Natl. Acad. Sci. U.S.A.* 75:2021 (1978).
2. T. Hunter and B. Sefton, Transforming gene product of Rous sarcoma virus phosphorylates tyrosine, *Proc. Natl. Acad. Sci. U.S.A.* 77:1311 (1980).
3. R. Doolittle, M.W. Hunkapiller, L.E. Hood, S.G. Devare, K.C., Robbins, S.A. Aaronson, and H.N. Antoniades, Simian sarcoma virus oncogene, v-*sis*, is derived from the gene (or genes) encoding a platelet-derived growth factor, *Science* 221:275 (1983); M. Waterfield, G.T. Scrace, N. Whittle, P. Stroobant, A. Johnsson, A. Wasteson, B. Westermark, C.H. Heldin, J.S. Huang, and T.F. Deuel, Platelet-derived growth factor is structurally related to the putative transforming protein p28sis of simian sarcoma virus, *Nature* 304:35 (1983).
4. J. Downward, Y. Yarden, E. Mayes, G. Scrace, N. Totty, P. Stockwell, A. Ullrich, J. Schlessinger, and M.D. Waterfield, Close similarity of epidermal growth factor receptor and v-*erb*-B oncogene protein sequences, *Nature* 307:521 (1984).
5. G. Carpenter and S. Cohen, Epidermal growth factor, *J. Biol. Chem.* 165:7709 (1990).
6. J.M. Bishop, Molecular themes in oncogenesis, *Cell* 64:235 (1991).
7. W.J. Pledger, C.D. Stiles, H.N. Antoniades, and C.D. Scher, Induction of DNA synthesis in BALB/c-3T3 cells by serum components: reevaluation of the commitment process, *Proc. Natl. Acad. Sci. U.S.A.* 74:4481 (1977); W.J. Pledger, C.D. Stiles, H.N. Antonaides, and C.D. Scher, An ordered sequence of events is required before BALB/c-3T3 cells become committed to DNA synthesis, *ibid.* 75:2839 (1978).
8. B. Westermark and C.H. Heldin, Similar action of platelet-derived growth factor and epidermal growth factor in the prereplicative phase of human fibroblasts suggests a common intracellular pathway, *J. Cell Physiol.* 124:43 (1985).
9. E.B. Leof, W. Wharton, J.J. Van Wyk, E.J. O'Keefe, and W.J. Pledger, Epidermal growth factor (EGF) and somatomedin C regulate G1 progression in competent BALB/c-3T3 cells, *Exp. Cell Res.* 141:107 (1982); E.B. Leof, J.J. Van Wyk, E.J. O'Keefe, and W.J. Pledger, Epidermal growth factor (EGF) is required only during the traverse of early G1 in PDGF stimulated density-arrested BALB/c-3T3 cells, *ibid.* 147:202 (1983).
10. D. Wexler, T.P. Fleming, P.P. Di Fiore, and S.A. Aaronson, unpublished observations.
11. A.B. Pardee, G1 events and regulation of cell proliferation, *Science* 246:603 (1989).
12. H.L. Moses, E.Y. Yang, and J.A. Pietenpol, TGF-beta stimulation and inhibition of cell proliferation: new mechanistic insights, *Cell* 63:245 (1990).
13. A.H. Wyllie, K.A. Rose, R.G. Morris, C.M. Steel, E. Foster, and D.A. Spandidos, Rodent fibroblast tumours expressing human *myc* and *ras* genes: growth, metastasis and endogenous oncogene expression, *Br. J. Cancer* 56:251 (1987); G.T. Williams, Programmed cell death: apoptosis and oncogenesis, *Cell* 65:1097 (1991).
14. D. Metcalf, The molecular control of cell division, differentiation commitment and maturation in haemopoietic cells, *Nature* 339:27 (1989).
15. F. Walker, N.A. Nicola, D. Metcalf, and A.N. Burgess, Hierarchical down-modulation of hemopoietic growth factor receptors, *Cell* 43:269 (1985); B.C. Gliniak and L.R. Rohrschneider, Expression of the M-CSF receptor is controlled posttranscriptionally by the dominant actions of GM-CSF or multi-CSF, *ibid.* 63:1073 (1990).
16. R.D. McKinnon, T. Matsui, M. Dubois-Dalcq, and S.A. Aaronson, FGF modulates the PDGF-driven pathway of oligodendrocyte development, *Neuron* 5:603 (1990).
17. J.H. Pierce, E. Di Marco, G.W. Cox, D. Lombardi, M. Ruggiero, L. Varesio, L.M. Wang, G.G. Choudhury, A.Y. Sakaguchi, and P.P. Di Fiore, Macrophage-colony-stimulating factor (CSF-1) induces proliferation, chemotaxis, and reversible monocytic differentiation in myeloid progenitor cells transfected with the human c-*fms*/CSF-1 receptor cDNA, *Proc. Natl. Acad. Sci. U.S.A.* 87:5613 (1990); L. Rohrschneider and D. Metcalf, Induction of macrophage colony-stimulating factor-

dependent growth and differentiation after introduction of the murine c-*fms* gene into FDC-P1 cells, *Mol. Cell. Biol.* 9:5081 (1989).

18. M.F. Roussel, T.J. Dull, C.W. Rettenmier, P. Ralph, A. Ullrich, and C.J. Sherr, Transforming potential of the c-*fms* proto-oncogene (CSF-1 receptor), *Nature* 325:549 (1987); J.H. Pierce, M. Ruggiero, T.P. Fleming, P.P. Di Fiore, J.S. Greenberger, L. Varticovski, J. Schlessinger, G. Rovera, and S.A. Aaronson, Signal transduction through the EGF receptor transfected in IL-3-dependent hematopoietic cells, *Science* 239:628 (1988); T. van Rüden and E.F. Wagner, Expression of functional human EGF receptor on murine bone marrow cells, *EMBO J.* 7:2749 (1988).
19. A. Ullrich and J. Schlessinger, Signal transduction by receptors with tyrosine kinase activity, *Cell* 61:203 (1990).
20. C.I. Bargmann, M.C. Hung, and R.A. Weinberg, Multiple independent activations of the *neu* oncogene by a point mutation altering the transmembrane domain of p185, *Cell* 45:649 (1986); O. Segatto, C.R. King, J.H. Pierce, P.P. Di Fiore, and S.A. Aaronson, Different structural alterations upregulate *in vitro* tyrosine kinase activity and transforming potency of the *erb*B-2 gene, *Mol. Cell. Biol.* 8:5570 (1988).
21. Y. Yarden and A. Ullrich, Growth factor receptor tyrosine kinases, *Annu. Rev. Biochem.* 57:443 (1988).
22. C.A. Koch, D. Anderson, M.F. Moran, C. Ellis, and T. Pawson, SH2 and SH3 domains: elements that control interactions of cytoplasmic signaling proteins, *Science* 252:668 (1991).
23. C.J. Sherr, C.W. Rettenmier, R. Sacca, M.F. Roussel, A.T. Look, and E.R. Stanley, The c-*fms* proto-oncogene product is related to the receptor for the mononuclear phagocyte growth factor, CSF-1, *Cell* 41:665 (1985); P. Besmer, J.E. Murphy, P.C. Goerge, F.H. Qiu, P.J. Bergold, L. Lederman, H.W. Snyder Jr., D. Brodeur, E.E. Zuckerman, and W.D. Hardy, A new acute transforming feline retrovirus and relationship of its oncogene v-*kit* with the protein kinase gene family, *Nature* 320:415 (1986); D.R. Smith, P.K. Vogt, and M.J. Hayman, The v-*sea* oncogene of avian erythroblastosis retrovirus S13: another member of the protein-tyrosine kinase gene family, *Proc. Natl. Acad. Sci. U.S.A.* 86:5291 (1989); H. Matshshime, L.H. Wang, and M. Shibuya, Human c-*ros*-1 gene homologous to the v-*ros* sequence of UR2 sarcoma virus encodes for a transmembrane receptorlike molecule, *Mol. Cell. Biol.* 6:3000 (1986).
24. C. Dickson, R. Deed, M. Dixon, and G. Peters, The structure and function of the int-2 oncogene, *Prog. Growth Fact. Res.* 1:123 (1989).
25. M. Taira, T. Yoshida, K. Miyagawa, H. Sakamoto, M. Terada, and T. Sugimura, cDNA sequence of human transforming gene *hst* and identification of the coding sequence required for transforming activity, *Proc. Natl. Acad. Sci. U.S.A.* 84:2980 (1987); P. Delli Bovi, A.M. Curatola, F.G. Kern, A. Greco, M. Ittmann, and C. Basilico, An oncogene isolated by transfection of Kaposi's sarcoma DNA encodes a growth factor that is a member of the FGF family, *Cell* 50:729 (1987); X. Zhan, B. Bates, X.G. Hu, and M. Goldfarb, The human FGF-5 oncogene encodes a novel protein related to fibroblast growth factors, *Mol. Cell. Biol.* 8:3487 (1988); A.L. Schechter, D.F. Stern, L. Vaidyanathan, S.J. Decker, J.A. Drebin, M.I. Greene, and R.A. Weinberg, The *neu* oncogene: an *erb*B-related gene encoding a 185,000-Mr tumour antigen, *Nature* 312:513 (1984); M. Dean, M. Park, M.M. Le Beau, T.S. Robins, M.O. Diaz, J.D. Rowley, D.G. Blair, and G.F. Vande Woude, The human *met* oncogene is related to the tyrosine kinase oncogenes, *ibid.* 318:385 (1985); D. Martin-Zanca, S.H. Hughes, and M. Barbacid, A human oncogene formed by the fusion of truncated tropomyosin and protein tyrosine kinase sequences, *ibid.* 319:743 (1986); M. Takahashi and G.M. Cooper, *ret* transforming gene encodes a fusion protein homologous to tyrosine kinases, *Mol. Cell. Biol.* 7:1378 (1987).
26. M. Shibuya, S. Yamaguchi, A. Yamane, T. Ikeda, A. Tojo, H. Matsushime, and M. Sato, Nucleotide sequence and expression of a novel human receptor-type tyrosine kinase gene (*flt*) closely related to the *fms* family, *Oncogene* 5:519 (1990); T. Matsui, M.A. Heidaran, T. Miki, N. Popescu, W.J. LaRochelle, M.H. Kraus, J.H. Pierce, and S.A. Aaronson, Isolation of anovel receptor cDNA establishes the existence of two PDGF receptor genes, *Science* 243:800 (1989); C.R. King, M.H. Kraus, and S.A. Aaronson, Amplification of a novel v-*erb*B-related gene in a human mammary

carcinoma, *ibid.* 229:974 (1985); K. Semba, N. Kamata, K. Toyoshima, and T. Yamamoto, A v-*erb*B-related protooncogene, c-*erb*B-2, is distinct from the c-*erb*B-1/epidermal growth factor-receptor gene and is amplified in a human salivary gland adenocarcinoma, *Proc. Natl. Acad. Sci. U.S.A.* 82:6497 (1985); P.L. Lee, D.E. Johnson, L.S. Cousens, V.A. Fried, and L.T. Williams, Purification and complementary DNA cloning of a receptor for basic fibroblast growth factor, *Science* 245:57 (1989); E.B. Pasquale and S.J. Singer, Identification of a developmentally regulated protein-tyrosine kinase by using anti-phosphotyrosine antibodies to screen a cDNA expression library, *Proc. Natl. Acad. Sci. U.S.A.* 86:5449 (1989); M. Ruta, R. Howk, G. Ricca, W. Drohan, M. Zabelshansky, G. Laureys, D.E. Barton, U. Francke, J. Schlessinger, and D. Givol, A novel protein tyrosine kinase gene whose expression is modulated during endothelial cell differentiation, *Oncogene* 3:9 (1988); Y. Hattori, H. Odagiri, H. Nakatani, K. Miyagawa, K. Naito, H. Sakamoto, O. Katoh, T. Yoshida, T. Sugimura, and M. Terada, K-sam, an amplified gene in stomach cancer, is a member of the heparin-binding growth factor receptor genes, *Proc. Natl. Acad. Sci. US.A.* 87:5983 (1990); S. Kornbluth, K.E. Paulson, and H. Hanafusa, Novel tyrosine kinase identified by phosphotyrosine antibody screening of cDNA libraries, *Mol. Cell. Biol.* 8:5541 (1988); E.B. Pasquale, A distinctive family of embryonic protein-tyrosine kinase receptors, *Proc. Natl. Acad. Sci. U.S.A.* 87:5812 (1990); T. Miki, T.P. Fleming, D.P. Bottaro, J.S. Rubin, D. Ron, and S.A. Aaronson, Expression cDNA cloning of the KGF receptor by creation of a transforming autocrine loop, *Science* 251:72 (1991); J. Partanen, T.P. Makela, E. Eerola, J. Korhonen, H. Hirvonen, L. Claesson-Welsh, and K. Alitalo, FGFR-4, a novel acidic fibroblast growth factor receptor with a distinct expression pattern, *EMBO J.* 10:1347 (1991); K. Keegan, D.E. Johnson, L.T. Williams, and M.J. Hayman, Isolation of an additional member of the fibroblast growth factor receptor family, FGFR-3, *Proc. Natl. Acad. Sci. U.S.A.* 88:3701 (1989); R. Klein, L.F. Parada, F. Coulier, and M. Barbacid, trkB, a novel tyrosine protein kinase receptor expressed during mouse neural development, *EMBO J.* 8:3701 (1989); H. Hirai, Y. Maru, K. Hagiwara, J. Nishida, and F. Takaku, A novel putative tyrosine kinase receptor encoded by the eph gene, Science 238:1717 (1987); V. Lhoták, P. Greer, K. Letwin, and T. Pawson, Characterization of elk, a grain-specific receptor tyrosine kinase, *Mol. Cell. Biol.* 11:2496 (1991); R.A. Lindberg and T. Hunter, cDNA cloning and characterization of eck, an epithelial cell receptor protein-tyrosine kinase in the eph/elk family of protein kinases, *ibid.* 10:6316 (1990).

27. M.H. Kraus, W. Issing, T. Miki, N.C. Popescu, and S.A. Aaronson, Isolation and characterization of *erb*B-3, a third member of the *erb*B/epidermal growth factor receptor family: evidence for overexpression in a subset of human mammary tumors, *Proc. Natl. Acad. Sci. U.S.A.* 86:9193 (1989).
28. A. Ullrich, J.R. Bell, E.Y. Chen, R. Herrera, L.M. Petruzzelli, T.J. Dull, A. Gray, L. Coussens, Y.C. Liao, and M. Tsubokawa, Human insulin receptor and its relationship to the tyrosine kinase family of oncogenes, *Nature* 313:756 (1987); Y. Yarden, J.A. Escobedo, W.J. Kuang, T.L. Yang-Feng, T.O. Daniel, P.M. Tremble, E.Y. Chen, M.E. Ando, R.N. Harkins, and U. Francke, Structure of the receptor for platelet-derived growth factor helps define a family of closely related growth factor receptors, *ibid.* 323:226 (1986); A. Ullrich, A. Gray, A.W. Tam, T. Yang-Feng, M. Tsubokawa, C. Collins, W. Henzel, T. Le Bon, S. Kathuria, and E. Chen, Insulin-like growth factor I receptor primary structure: comparison with insulin receptor suggests structural determinants that define functional specificity, *EMBO J.* 5:2503 (1986).
29. S.A. Aaronson and S.R. Tronick, Growth factors, *in*: Cancer Medicine, J.F. Holland III, R.C.. Bast Jr., D.W. Kufe, D.L. Morton, eds., Lea & Febiger, Philadelphia (1991).
30. D.P. Bottaro, J.S. Rubin, D.L. Faletto, A.M.-L. Chan, T.E. Kmiecik, G.F. Vande Woude, and S.A. Aaronson, Identification of the hepatocyte growth factor receptor as the c-*met* proto-oncogene product, *Science* 251:802 (1991); L. Naldini, E. Vigna, R.P. Narsimhan, G. Gaudino, R. Zarnegar, G.K. Michalopoulos, and P.M. Comoglio, Hepatocyte growth factor (HGF) stimulates the tyrosine kinase activity of the receptor encoded by the proto-oncogene c-MET, *Oncogene* 6:501 (1991).

31. B.L. Hempstead, D. Martin-Zanca, D.R. Kaplan, L.F. Parada, M.V. Chao, High affinity NGF binding requires coexpression of the *trk* proto-oncogene and the low affinity NGF receptor, *Nature* 350:678 (1991); A.R. Nebreda, D. Martin-Zanca, D.R. Kaplan, L.F. Parada, and E. Santos, Induction by NGF of meiotic maturation of *Xenopus oocytes* expressing the *trk* proto-oncogene product, *Science* 252:558 (1991); R. Klein, S.Q. Jing, V. Nanduri, E. O'Rourke, and M. Barbacid, The *trk* proto-oncogene encodes a receptor for nerve growth factor, *Cell* 65:189 (1991).
32. D. Soppet, E. Escandon, J. Maragos, D.S. Middlemas, S.W. Reid, J. Blair, L.E. Burton, B.R. Stanton, D.R. Kaplan, and T. Hunter, The neurotrophic factors, brain-derived neurotrophic factor and neurotrophin-3 are ligands for the trkB tyrosine kinase receptor, *Cell* 65:895 (1991).
33. J. Meisenhelder, P.-G. Suh, S.G. Rhee, and T. Hunter, Phospholipase C-gamma is a substrate for the PDGF and EGF receptor protein-tyrosine kinases *in vivo* and *in vitro*, *Cell* 57:1109 (1989); M.I. Wahl, N.E. Olashaw, S. Nishibe, S.G. Rhee, W.J. Pledger, and G. Carpenter, Platelet-derived growth factor induces rapid and sustained tyrosine phosphorylation of phospholipase C-gamma in quiescent BALB/c 3T3 cells, *Mol. Cell. Biol.* 9:2934 (1989).
34. D.R. Kaplan, M. Whitman, B. Schaffhausen, D.C. Pallas, M. White, L. Cantley, and T.M. Roberts, Common elements in growth factor stimulation and oncogenic transformation: 85 kd phosphoprotein and phosphatidylinositol kinase activity, *Cell* 50:1021 (1987).
35. C.J. Molloy, D.P. Bottaro, T.P. Fleming, M.S. Marshall, J.B. Gibbs, and S.A. Aaronson, PDGF induction of tyrosine phosphorylation of GTPase activating protein, *Nature* 342:711 (1989); D.R. Kaplan, D.K. Morrison, G. Wong, F. McCormick, and L.T. Williams, PDGF beta-receptor stimulates tyrosine phosphorylation of GAP and association of GAP with a signaling complex, *Cell* 61:125 (1990); A. Kazlauskas, C. Ellis, T. Pawson, J.A. Cooper, Binding of GAP to activated PDGF receptors, *Science* 247:1578 (1990).
36. R. Ralston and J.M. Bishop, The product of the protooncogene c-*src* is modified during the cellular response to platelet-derived growth factor, *Proc. Natl. Acad. Sci. U.S.A.* 82:7845 (1985); R.M. Kypta, Y. Goldberg, E.T. Ulug, and S.A. Courtneidge, Association between the PDGF receptor and members of the *src* family of tyrosine kinases, *Cell* 62:481 (1990).
37. D.K. Morrison, D.R. Kaplan, U.R. Rapp, and T.M. Roberts, Signal transduction from membrane to cytoplasm: growth factors and membrane-bound oncogene products increase Raf-1 phosphorylation and associated protein kinase activity, *Proc. Natl. Acad. Sci. U.S.A.* 85:8855 (1988); D.K. Morrison, D.R. Kaplan, J.A. Escobedo, U.R. Rapp, T.M. Roberts, and L.T. Williams, Direct activation of the serine/threonine kinase activity of Raf-1 through tyrosine phosphorylation by the PDSGF beta receptor, *Cell* 58:649 (1989).
38. A. Berridge and R.F. Irvine, Inositol phosphates and cell signaling, *Nature* 341:197 (1989); U. Kikkawa, A. Kishimoto, and Y. Nishizuka, The protein kinase C family: heterogeneity and its implications, *Annu. Rev. Biochem.* 58:31 (1989).
39. S. Nishibe, M.I. Wahl, S.M. Hernandez-Sotomayor, N.K. Tonks, S.G. Rhee, and G. Carpenter, Increase of the catalytic activity of phospholipase C-gamma 1 by tyrosine phosphorylation, *Science* 250:1253 (1990); P.J. Goldschmidt-Clermont, J.W. Kim, L.M. Machesky, S.G. Rhee, and T.D. Pollard, Regulation of phospholipase C-gamma 1 by profilin and tyrosine phosphorylation, *ibid.* 251:1231 (1991).
40. FL.C. Cantley, K.R. Auger, C. Carpenter, B. Duckworth, A. Graziani, R. Kapeller, and S. Soltoff, Oncogenes and signal transduction, *Cell* 64:281 (1991).
41. J.A. Escobedo, S. Navankasattusas, W.M. Kavanaugh, D. Milfay, V.A. Fried, and L.T. Williams, cDNA cloning of a novel 85 kd protein that has SH2 domains and regulates binding of PI3-kinase to the PDGF beta receptor, *Cell* 65:75 (1991); E.Y. Skolnik, B. Margolis, M. Mohammadi, E. Lowenstein, R. Fischer, A. Drepps, A. Ullrich, and J. Schlessinger, Cloning of PI3 kinase-associated p85 utilizing a novel method for expression/cloning of target proteins for receptor tyrosine kinases, *ibid.* 65:83 (1991); M. Otsu, I. Hiles, I. Gout, M.J. Fry, F. Ruiz-Larrea, G. Panayotou, A. Thompson, R. Dhand, J. Hsuan, and N. Totty, Characterization of two 85 kd proteins that associate with receptor tyrosine kinases, middle-T/pp60c-*src* complexes, and PI3-kinase, *ibid.* 65:91 (1991).

42. M. Whitman, D.R. Kaplan, B. Schaffransen, L. Cantley, and T.M. Roberts, Association of phosphatidylinositol kinase activity with polyoma middle-T competent for transformation, *Nature* 315:239 (1985); D.R. Kaplan, M. Whitman, B. Schaffhausen, L. Raptis, R.L. Garcea, D. Pallas, T.M. Roberts, and L. Cantley, Phosphatidylinositol metabolism and polyoma-mediated transformation, *Proc. Natl. Acad. Sci. U.S.A.* 83:3624 (1986); S.A. Courtneidge and A. Heber, An 81 kd protein complexed with middle T antigen and pp60c-*src*: a possible phosphatidylinositol kinase, *Cell* 50:1031 (1987).
43. Y. Fukui and H. Hanafusa, Phosphatidylinositol kinase activity associates with viral p60src protein, *Mol. Cell. Biol.* 9:1651 (1989); L. Varticovski, Q. Daley, P. Jackson, D. Baltimore, and L.C. Cantley, Activation of phosphatidylinositol 3-kinase in cells expressing *abl* oncogene variants, *ibid.* 11:1107 (1991).
44. F. McCormick, *ras* GTPase activating protein: signal transmitter and signal terminator, *Cell* 56:5 (1989).
45. M. Barbacid, *ras* genes, *Annu. Rev. Biochem.* 56:779 (1987).
46. D.W. Stacey and H.F. Kung, Transformation of NIH 3T3 cells by microinjection of Ha-*ras* p21 protein, *Nature* 310:508 (1984).
47. K. Zhang, J.E. DeClue, W.C. Vass, A.G. Papageorge, F. McCormick, and D.R. Lowy, Suppression of c-*ras* transformation by GTPase-activating protein, *Nature* 346:754 (1990); K. Tanaka, K. Matsumoto, and E.A. Toh, IRA1, an inhibitory regulator of the RAS-cyclic AMP pathway in *Saccharomyces cerevisiae*, *Mol. Cell. Biol.* 9:757 (1989); R. Ballester, T. Michaeli, K. Ferguson, H.P. Xu, F. McCormick, and M. Wigler, Genetic analysis of mammalian GAP expressed in yeast, *Cell* 59:681 (1989).
48. A. Yatani, K. Okabe, P. Polakis, R. Halenbeck, F. McCormick, and A.M. Brown, *ras* p21 and GAP inhibit coupling of muscarinic receptors to atrial K+ channels, *Cell* 61:769 (1990).
49. A.H. Bouton, S.B. Kanner, R.R. Vines, H.C. Wang, J.B. Gibbs, and J.T. Parsons, Transformation by pp60src or stimulation of cells with epidermal growth factor induces the stable association of tyrosine-phosphorylated cellular proteins with GTPase-activating protein, *Mol. Cell. Biol.* 11:945 (1991).
50. M.F. Moran, P. Polakis, F. McCormick, T. Pawson, and C. Ellis, Protein-tyrosine kinases regulate the phosphorylation, protein interactions, subcellular distribution and activity of p21ras GTPase-activating protein, *Mol. Cell. Biol.* 11:1807 (1991).
51. T. Satoh, M. Endo, M. Nakafuku, S. Nakamura, Y. Kasiro, Platelet-derived growth factor stimulates formation of active p21ras GTP complex in Swiss mouse 3T3 cells, *Proc. Natl. Acad. Sci. U.S.A.* 87:5993 (1991); J.B. Gibbs, M.S. Marshall, E.M. Scolnik, R.A. Dixon, and U.S. Vogel, Modulation of guanine nucleotides bound to Ras in NIH3T3 cells by oncogenes, growth factors and the GTPase activating protein (GAP), *J. Biol. Chem.* 265:20437 (1990).
52. S. Powers, E. Gonzales, T. Christensen, J. Cubert, and D. Broek, Functional cloning of BUD5, a CDC25-related gene from *S. cerevisiae* that can suppress a dominant-negative RAS2 mutant, *Cell* 65:12125 (1991); J. Chant, K. Corrado, J. Pringle, and I. Herskowitz, Yeast BUD5, encoding a putative GDP-GTP exchange factor, is necessary for bud site selection and interacts with bud formation gene BEM1, *ibid.* 65:1231 (1991); S. Jones, M.L. Vignais, and J.R. Broach, The CDC25 protein of *Saccharomyces cerevisiae* promotes exchange of guanine nucleotides bound to *ras*, *Mol. Cell. Biol.* 11:2641 (1991).
53. Y.K. Huang, H.F. Kung, T. Kamata, Purification of a factor capable of stimulating the guanine nucleotide exchange reaction of ras proteins and its effect on ras-related small molecular mass G proteins, *Proc. Natl. Acad. Sci. U.S.A.* 87:8008 (1990); A. Wolfman and I.G. Macara, A cytosolic protein catalyzes the release of GDP from p21ras, *Science* 248:67 (1990); J. Downward, R. Riehl, L. Wu, and R.A. Weinberg, Identification of a nucleotide exchange-promoting activity for p21ras, *Proc. Natl. Acad. Sci. U.S.A.* 87:5998 (1990).
54. J.A. Cooper, The SRC family of protein tyrosine kinases*in*: Peptides and Protein Phosphorylation, B. Kenpard and P.F. Alewood, eds. CRC Press, Inc., Boca Raton, (1990).

55. U.R. Rapp, G. Heidecker, M. Huleihel, J.L. Cleveland, W.C. Choi, T. Pawson, J.N. Ihle, and W.B. Anderson, raf family serine/threonine protein kinases in mitogen signal transduction, *Cold Spring Harbor Symposium Quant. Biol.* 53:173 (1988).
56. P. Li, K. Wood, H. Mamon, W. Haser, and T. Roberts, Raf-1: a kinase currently without a cause but not lacking in effects, *Cell* 64:479 (1991).
57. J.P. Falco, W.G. Taylor, P.P. Di Fiore, B.E. Weissman, and S.A. Aaronson, Interactions of growth factors and retroviral oncogenes with mitogenic signal transduction pathways of Balb/MK keratinocytes, *Oncogene* 2:573 (1988).
58. A. Balmain and I.B. Pragnell, Mouse skin carcinomas induced *in vivo* by chemical carcinogens have a transforming Harvey-*ras* oncogene, *Nature* 303:72 (1983); B. Bailleul, M.A. Surani, S. White, S.C. Barton, K. Brown, M. Blessing, J. Jorcano, and A. Balmain, Skin hyperkeratosis and papilloma formation in transgenic mice expressing a *ras* oncogene from a suprabasal keratin promoter, *Cell* 62:697 (1990).
59. S. Alema, P. Cassalbore, E. Agostini, F. Tabo, Differentiation of PC12 phaeochromocytoma cells induced by v-*src* oncogene, *Nature* 316:557 (1985); M. Noda, M. Ko, A. Ogura, D.G. Liu, T. Amano, T. Takano, and Y. Ikawa, Sarcoma viruses carrying *ras* oncogenes induce differentiation-associated properties in a neuronal cell line, *ibid.* 318:73 (1985); I. Guerrero, H. Wong, A. Pellicer, and D.E. Burstein, Activated N-*ras* gene induces neuronal differentiation of PC12 rat pheochromocytoma cells, *J. Cell Physiol.* 129:71 (1986).
60. S. Seremetis, G. Inghirami, D. Ferrero, D.W. Newcomb, D.M. Knowles, G.P. Dotto, and R. Dalla-Favera, Transformation and plasmacytoid differentiation of EBV-infected human B lymphoblasts by *ras* oncogenes, *Science* 243:660 (1989).
61. D. Cosman, S.D. Lyman, R.L. Idzerda, M.P. Beckmann, L.S. Park, R.G. Goodwin, and C.J. March, A new cytokine receptor superfamily, *Trends Biochem. Sci.* 15:265 (1990).
62. M. Tsudo, R.W. Kozak, C.K. Goldman, and J.A. Waldmann, Demonstration of a non-Tac peptide that binds interleukin 2: a potential participant in a multichain interleukin 2 receptor complex, *Proc. Natl. Acad. Sci. U.S.A.* 84:9694 (1986); M. Hibi, M. Murakami, M. Saito, T. Hirano, T. Taga, and T. Kishimoto, Molecular Cloning and expression of an IL-6 signal transducer, gp130, *Cell* 63:1149 (1990).
63. A. Morla, J. Schreur, A. Miyajama, J. Wang, *Mol. Cell. Biol.* 8:2214 (1988); Y. Konakura et al *Blood* 76:706 (1990); R. Isfort, R. Hunn, R. Frackelton, and J. Ihle, *J. Biol. Chem.* 263:19203 (1988).
64. T. Satoh, M. Nakafuku, A. Miyajima, Y. Kaziro, Involvement of *ras* p21 protein in signal transduction pathways from interleukin 2, interleukin 3, and granulocyte/macrophage colony-stimulating factor, but not from interleukin 4, *Proc. Natl. Acad. Sci. U.S.A.* 88:3314 (1991).
65. M. Hatakeyama, T. Kono, N. Kobayashi, A. Kawahara, S.D. Levin, R.M. Perlmutter, and T. Taniguchi, Interaction of the IL-2 receptor with the *src*-family kinase p56lck: identification of novel intermolecular association, *Science* 252:1523 (1991).
66. A. Yoshinura, G. Longmore, and H. Lodish, Point mutation in the exoplasmic domain of the erythropoietin receptor resulting in hormone-independent activation and tumorigenicity, *Nature* 348:647 (1990).
67. S.K. Ruscetti, N. Janesch, A. Chakroborti, S.T. Sawyer, and W.D. Hankins, Friend spleen focus-forming virus induces factor independence in an erythropoietin-dependent erythroleukemia cell line, *J. Virol.* 63:1057 (1990); J.P. Li, A. D'Andrea, H. Lodish, and D. Baltimore, Activation of cell growth by binding of Friend spleen focus-forming virus gp55 glycoprotein to the erythropoietin receptor, *Nature* 343:762 (1990).
68. F. Wong-Staal and R. Gallo, Human T-lymphotropic retroviruses, *Nature* 317:395 (1985).
69. H.R. Bourne and A.L. DeFranco, Signal transduction and intracellular messengers *in*: Oncogenes and the Molecular Origins of Cancer, R.E. Weinberg, ed. Cold Spring Harbor Laboratory, Cold Spring Harbor (1989).
70. K.R. Stratton, P.F. Worley, R.L. Huganir, and J.M. Baraban, Muscarinic agonists and phorbol esters increase tyrosine phosphorylation of a 40-kilodalton protein in hippocampal slices, *Proc. Natl.*

Acad. Sci. U.S.A. 86:2498 (1989); J. Zachary, J. Gil, W. Lehman, and J. Sinnett-Smith, Bombesin, vasopressin and endothelin rapidly stimulate tyrosine phosphorylation in intact Swiss 3T3 cells, *ibid.* 88:4577 (1991); T. Force, J.M. Kyriakis, J. Avruch, and J.V. Bonventre, Endothelin, vasopressin and angiotensin II enhance tyrosine phosphorylation by protein kinase C-dependent and -independent pathways in glomerular mesangial cells, *J. Biol. Chem.* 266:6650 (1991); L.M.T. Leeb-Lundberg and S.-H. Song, Bradykinin and bombesin rapidly stimulate tyrosine phosphorylation of a 120-kDa group of proteins in Swiss 3T3 cells, *ibid.* 266:7746 (1991); S.J. Shattil and J.S. Brugge, Protein tyrosine phosphorylation and the adhesive functions of platelets, *Curr. Opin. Cell Biol.*. 3:869 (1991).

71. B.F. O'Dowd, R.J. Lefkowitz, and M.G. Caron, Structure of the adrenergic and related receptors, *Annu. Rev. Neurosci.* 12:67 (1989); J. Ramachandran, E.G. Peralta, A. Ashkenazi, J.W. Winslow, and D.J. Capon, The structural and functional interrelationships of muscarinic acetylcholine receptor subtypes, *Bioessays* 10:54 (1989); H.Y. Lin, E.H. Kaji G.K. Winkel, H.E. Ives, and H.F. Lodish, Cloning and functional expression of a vascular smooth muscle endothelin 1 receptor, *Proc. Natl. Acad. Sci. U.S.A.* 88:3185 (1991).
72. T.R. Jackson, L.A. Blair, J. Marshall, M. Goedert, and M.R. Hanley, The mas oncogene encodes an angiotensin receptor, *Nature* 335:440 (1988).
73. D. Julius, T.J. Livelli, T.M. Jessell, and R. Axel, Ectopic expression of the serotonin 1c receptor and the triggering of malignant transformation, *Science* 244:1057 (1989).
74. F. Cuttitta, D.N. Carney, J. Mulshine, T.W. Moody, J. Fedorko, A. Fischler, and J.D. Minna, Bombesin-like peptides can function as autocrine growth factors in human small-cell lung cancer, *Nature* 316:823 (1985).
75. J.L. Mulshine, I. Avis, A.M. Treston, C. Mobley, P. Kaspryzyk, J.A. Carrasquillo, S.M. Larson, Y. Nakanishi, B. Merchant, and J.D. Minna, Clinical use of a monoclonal antibody to bombesin-like peptide in patients with lung cancer, *Ann. N.Y. Acad. Sci.* 547:360 (1988).
76. C.A. Landis, S.B. Masters, A. Spada, A.M. Pace, H.R. Bourne, and L. Vallar, GTPase inhibiting mutations activate the alpha chain of Gs and stimulate adenylyl cyclase in human pituitary tumours, *Nature* 340:692 (1989); J. Lyons, C.A. Landis, G. Harsh, L. Vallar, K. Gruenwald, H. Feichtinger, Q.Y. Duh, O.H. Clark, E. Kawasaki, and H.R. Bourne, Two G protein oncogenes in human endocrine tumors, *Science* 249:655 (1990).
77. J.S. Gutkind, E.A. Novotny, M.R. Brann, and K.C. Robbins, Muscarinic acetylcholine receptor subtypes as agonist-dependent oncogenes, *Proc. Natl. Acad. Sci. U.S.A.* 88:4703 (1991).
78. T. Miki, T.P. Fleming, M. Crescenzi, C.J. Molloy, S.B. Blam, S.H. Reynolds, and S.A. Aaronson, Development of a highly efficient expression cDNA cloning system: application to oncogene isolation, *Proc. Natl Acad. Sci. U.S.A.* 88:5167 (1991).
79. M.P. Strathmann and M.I. Simon, Ga12 and Ga13 subunits define a fourth class of G protein a subunits, *Proc. Natl. Acad. Sci. U.S.A.* 88:5582 (1991).
80. A.M.-L. Chan, T.P. Fleming, E.S. McGovern, M. Chedid, T. Miki, and S.A. Aaronson, Expression cDNA cloning of a transforming gene encoding the wild-type Ga12 gene product, *Mol. Cell. Biol.* 13:762 (1993).
81. T. Miki, D.P. Bottaro, T.P. Fleming, C.L. Smith, W.H. Burgess, A.M.-L. Chan, and S.A. Aaronson, Determination of ligand-binding specificity by alternative splicing: two distinct growth factor receptors encoded by a single gene, *Proc. Natl. Acad. Sci. U.S.A.* 89:246 (1992).
82. E. Solomon, J. Borrow, and A.D. Goddard, Chromosome aberrations and cancer, *Science* 254:1153 (1991).
83. R.A. Weinberg, Tumor suppressor genes, *Science* 254:1138 (1991).
84. C. Shih, B.-Z. Shilo, M.P. Goldfarb, A. Dannenberg, and R.A. Weinberg, Passage of phenotypes of chemically transformed cells via transfection of DNA and chromatin, *Proc. Natl. Acad. Sci. U.S.A.* 76:5714 (1979).
85. A.M.-L. Chan, unpublished observations.

TUMOR SUPPRESSOR GENES WHICH ENCODE TRANSCRIPTIONAL REPRESSORS: STUDIES ON THE EGR AND WILMS' TUMOR (WT1) GENE PRODUCTS

Frank J. Rauscher, III

The Wistar Institute of Anatomy and Biology
3601 Spruce Street
Philadelphia, PA 19104 U.S.A.

ONCOGENES AND SIGNAL TRANSDUCTION

Over the past 15 years a revolution has occurred in our understanding of the cellular and molecular bases of neoplastic transformation and subsequent tumor formation. The recognition that neoplastic cell growth is a stable, heritable cellular phenotype led directly to the cloning of specific genetic loci from human tumor DNAs (termed oncogenes) which could convert a normal cell into a malignant cell (reviewed in reference 1). The remarkable finding that many oncogenic murine and avian retroviruses carried mutated versions of these cellular oncogenes gave further support for the oncogene hypothesis.[1] Support for the relevance of oncogenes to human disease has come from the realization that many of these cancer-genes are "activated' (either by mutation, over-expression, or inappropriate expression) in many human tumor specimens.[2]

Elucidating the cellular bases for oncogene action and the roles of potential oncogenes in normal ontogeny, growth and differentiation processes has been a field of intense research. Oncogene research has essentially spawned and established a new field which can be broadly termed "cellular signal transduction." The establishment of this field has come from the realization that almost all oncogenic proteins appear to function as information transducers within or between cells. That is, the proteins encoded by oncogenes perform the biochemical functions necessary for a proliferative or differentiative signal received at the cell membrane to eventually arrive at the nucleus where it results in a modification of gene expression. Examples of signal transduction roles for oncogenes (shown schematically in Figure 1) include secreted growth factors (c-*sis*), growth-factor receptors (c-*erb*), receptor-linked guanine-nucleotide binding proteins (c-*ras*), serine/threonine and tyrosine kinases or phosphatases (c-*src*, c-*raf*, c-*mos*) and finally, nuclear localized proteins which function as transcription factors (c-*myc*, c-*fos*, c-*jun*, c-*myb*). Thus, information transfer from the extracellular environment to the nucleus is accomplished by a cascade of events in a highly compartmentalized but integrated system that is carried out by the products of cellular proto

Advances in Nutrition and Cancer, Edited by
V. Zappia *et al.*, Plenum Press, New York, 1993

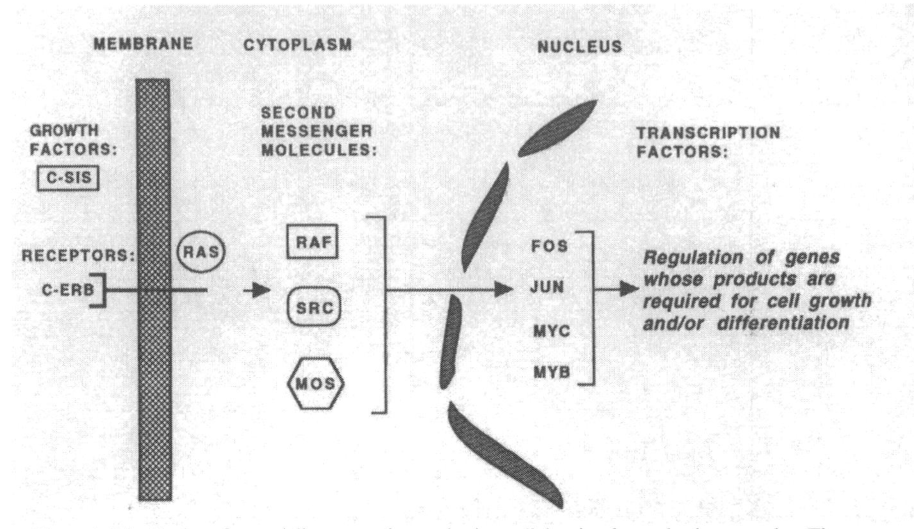

Figure 1. Distribution of potentially oncogenic proteins in a cellular signal transduction cascade. (The arrows do not indicate that direct, functional interactions occur between these proteins in the pathway)

oncogenes. A block in this process at almost any step in the cascade can result in aberrant transduction of signals ultimately resulting in uncontrolled cell growth we see as tumor formation. Clearly, this signal transduction system must be highly regulated, fairly redundant in a molecular sense, and flexible enough to respond very rapidly to changes in the extracellular milieu.

TRANSCRIPTION FACTORS AS SIGNAL TRANSDUCTION MOLECULES

One of the major questions currently facing the oncogenologist is: What is the nature of the "output" from this signal transduction system and how does this output determine whether the cell will stop dividing and differentiate or continue proliferating? It has been become apparent that each of these cellular responses to a signal (either proliferation or differentiation) requires activation of a distinct program of gene expression in the nucleus. The set of genes which are activated or repressed by the extracellular stimulus, encodes the proteins which will ultimately change the proliferative or differentiative phenotype of the cell.

Since a change in gene expression is required to establish a new cellular phenotype, much effort has been directed at determining the role of transcription factors in modulating gene expression in response to growth factors.[3] Transcription factors which bind DNA sequence-specifically and regulate the initiation of transcription can be thought of as the ultimate downstream receptors for signals which are initially received at the cell membrane. Transcription factors can rapidly couple short-term signals received at the cell membrane to long-term adaptive changes in cell phenotype (i.e., exit from the cell-cycle and/or differentiation) by coordinately regulating the expression of target genes.[4] It is crucial that we identify the transcription factors in the nucleus which positively and negatively influence cell-growth and ultimately identify the target genes which are regulated by these factors.

REGULATION OF GENE EXPRESSION BY THE EARLY GROWTH RESPONSE (EGR) FAMILY OF TRANSCRIPTION FACTORS

The study of the EGR family of zinc finger-containing transcription factors and the product of the WT1 Wilms' tumor suppressor locus has provided a molecular paradigm for how positive and negative regulation of cell growth can be elicited through regulation of a common set of target genes.

The EGR family of proteins is comprised of four members (Figure 2) (reviewed in reference 5). EGR-1 is the prototype of the family and was initially cloned from a library of growth factor induced cDNAs.[6] The gene encodes a 533 amino acid protein which contains three zinc fingers of the C_2H_2 class and a serine-glutamine-proline-rich N-terminus. The EGR-1 protein binds to the DNA sequence 5'-GCGGGGGCG-3' and functions as an activator of transcription when bound to that sequence in a target gene in vivo.[5] EGR-2, -3, and -4 were isolated as growth-factor-induced genes or by low stringency screening of cDNA libraries using an EGR-1 cDNA as a probe.[5] Each gene encodes a protein with three zinc fingers which share greater than 90% amino acid sequence homology among the family with EGR-1, however, there is little sequence homology outside of the zinc finger region (Figure 2). As expected, EGR-2, -3 and -4 also bind to the EGR consensus sequence and function as activators of transcription. Each of the EGR genes are mitogen- and growth-factor-inducible "immediate-early" genes.[5] These genes are induced within five minutes of growth-factor addition to quiescent cells. The EGR proteins rapidly accumulate in the nucleus where it is thought they regulate a set of target genes required for entry into the cell cycle and subsequent DNA synthesis. Thus, the EGR family of genes participate in signal transduction processes in the nucleus by rapidly responding to extracellular stimuli and regulating expression of specific target genes.

Clearly, this rapidly-induced transcription factor system in the nucleus must be just as rapidly down-regulated such that the proliferative signal is turned off. The down-regulation of EGR family gene transcription and protein function has been shown to occur at many levels. Transcription of the genes are rapidly down-regulated, the mRNAs are processed and rapidly degraded and the proteins themselves are rapidly turned over.[5] Recently, a new insight into down-regulation of the EGR response has come from a biochemical analysis of the WT1 Wilms' tumor protein. As described below, WT1 is also a zinc finger protein which binds to the EGR consensus sequence and functions as a repressor of transcription.

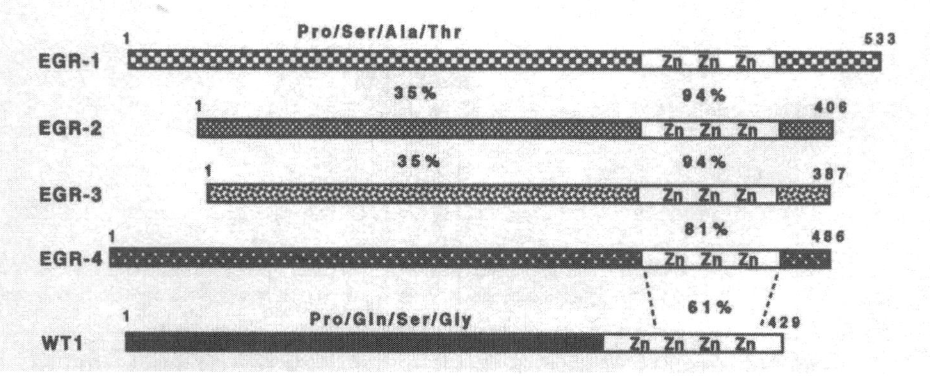

Figure 2. The Early Growth Response (EGR) gene family. The number of amino acids contained in each protein, the percent sequence homology of each protein, and the relative location of the three zinc-fingers are indicated.

WILMS' TUMOR OF THE KIDNEY: GENETICS AND BIOLOGY

Wilms' tumor is a pediatric embryonal cell malignancy of the kidney. It occurs at a frequency of about 1 in 10,000 live-births and is observed in both heritable and sporadic forms, much the same as retinoblastoma (reviewed in reference 7). Knudson and Strong[8] proposed that development of Wilms' tumor and retinoblastoma required two genetic hits (or mutations) and, therefore, that a recessive oncogene(s) may be involved. The association of Wilms' tumor with aniridia (a developmental defect of the iris) allowed provisional mapping of a Wilms' tumor susceptibility locus to chromosome 11, bands 13-15.[7] Through an elegant series of positional cloning and chromosome jumping experiments, a gene (WT1) was isolated from the 11p13 region which had the characteristics of the disease susceptibility locus.[9,10] The WT1 gene is mutated in 20-30% of sporadic Wilms' tumors and in 100% of Wilms' tumor associated with the Denys-Drash Syndrome.[11] The gene is expressed primarily in the differentiating mesenchyme of embryonic kidney during formation of the primitive glomerulus.[12] The expression pattern suggests the WT1 protein shuts off cell growth processes and establishes and maintains a differentiated phenotype in the podocytes of the kidney. Loss of WT1 function (due to point mutation or deletion of the chromosomal locus) apparently leads to uncontrolled proliferation of mesenchymal cells resulting in the development of a Wilms' tumor.

The WT1 protein contains four zinc fingers of the C_2H_2 class and a proline-glutamine-glycine-rich N-terminus.[9,10] Zinc fingers two, three and four share ~60% amino acid sequence homology with the EGR family zinc fingers (Figure 2). To determine the DNA binding site recognized by the WT1 protein, we used a library of degenerate oligonucleotides for affinity selection with recombinant WT1 protein.[13] The binding sites we obtained were very GC-rich and resembled the EGR consensus sequence. Subsequently, we demonstrated that EGR-1 and WT1 bind to the same target sequence.[13,14] Recently, a crystallographic study of the EGR-1 protein bound to its consensus sequence has provided a structural basis for recognition of DNA by EGR-1 and WT1.[15] This study identified the amino acids in EGR-1 which directly contact DNA, as either arginine-guanine or histidine-guanine contacts. As shown schematically in Figure 3, WT1 contains the exact same arginine and histidine residues in the critical DNA contact positions. Thus, the protein encoded by a tumor suppressor gene, (a putative negative regulator of cell growth) WT1 binds to the same DNA consensus sequence recognized by a growth-factor-inducible, mitogen-stimulated gene, EGR-1.

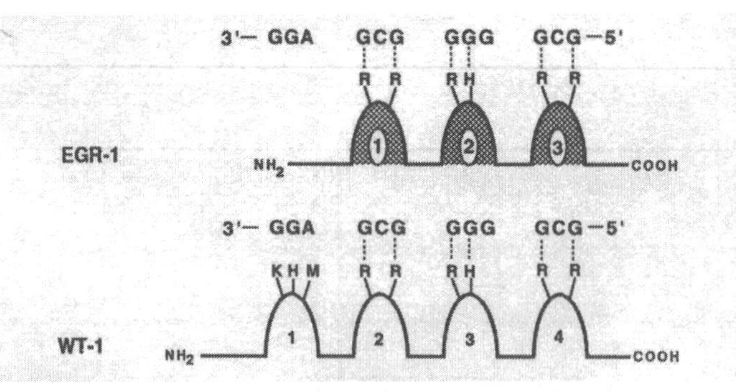

Figure 3. The relative orientation of the EGR1 and WT1 zinc fingers when bound to the EGR consensus DNA sequence. Only one strand of the EGR consensus duplex is shown. The location of the arginine (R) and histidine (H) residues which contact the nucleotide bases are shown. Note the presence of the additional zinc finger in WT1. The important contact residues and the nucleotides recognized by this additional finger are not yet defined.

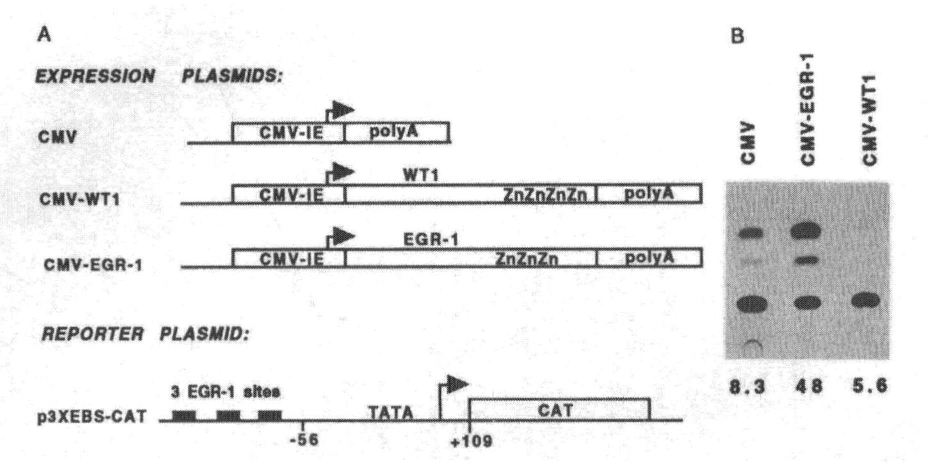

Figure 4. Activation and repression of transcription by EGR-1 and WT1 respectively. A. Expression vectors containing full-length cDNAs for EGR-1 or WT1 were co-transfected with a reporter vector which contains their EGR binding sites upstream of a minimal promoter driving the chloramphenicol-acetyl transferase gene (CAT). B. CAT assays of transfected cells. The basal CAT activity of the reporter vector is shown in the left lane. Co-transfection with CMV-EGR-1 activated transcription (middle lane) whereas co-transfection with CMV-WT1 repressed transcription (right lane). The relative percent conversion is shown at the bottom of each lane. (Part B was reprinted from *Science* (reference 16) with permission)

REPRESSION OF TRANSCRIPTION BY THE WT1 WILMS' TUMOR GENE PRODUCT

Since both EGR-1 and WT1 bind to the same target sequence, we sought to determine the transcriptional regulatory potential of WT1. We performed transient co-transfection assays using plasmid vectors which express either WT1 or EGR-1 proteins and reporter plasmids which contain synthetic EGR binding sites (Figure 4). As expected, EGR-1 func-

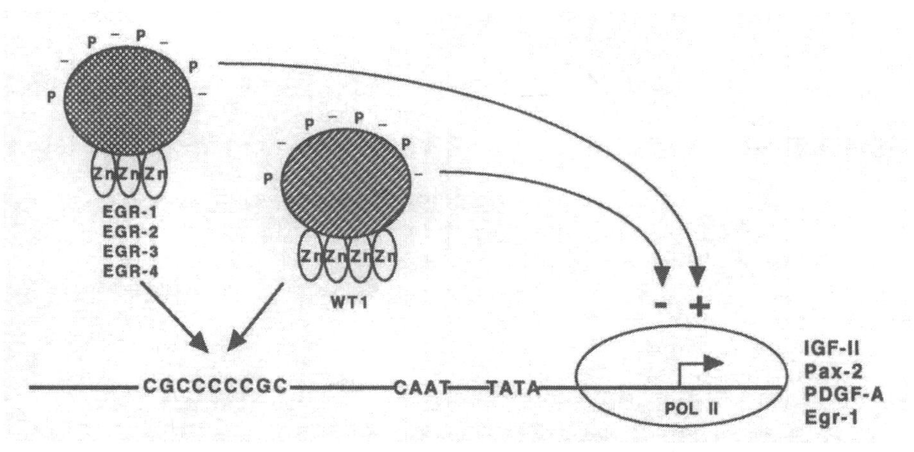

Figure 5. A model for the regulation of transcription by WT1 and EGR family proteins. Each protein binds the same *cis*-acting regulatory element yet has opposite effects on transcription. The ultimate output from this set of target genes should depend on the relative concentrations and activities of WT1 and EGRs in the cell. At right is a list of target genes which have been shown to be regulated by WT1 and EGR-1.

tions as an activator of transcription. however, WT1 functions as a potent repressor of transcription from the same reporter plasmids.[16,17] A mutagenesis and truncation analysis of WT1 demonstrated that both the zinc finger region of WT1 and the glutamine-proline-glycine-rich N-terminus of WT1 are required for transcriptional repression.[18]

Thus EGR-1 and WT1 display opposite effects on the transcription of a target gene containing EGR binding sites (Figure 5): EGR-1 activates transcription, WT1 represses transcription. Clearly the relative ratio of WT1 to EGR-1 in a cell at a particular time in development will govern what the "output" is from the set of target genes containing EGR binding sites. A disruption in this ratio very likely occurs when WT1 is mutated or deleted (as is observed in cases of Wilms' tumor). The unrestrained growth of mesenchymal cells (characteristic of Wilms' tumor) in the developing kidney may be due to unrestrained activation of target genes regulated by the EGR family, thereby producing a sustained proliferative signal.

SUMMARY

The study of the WT1 and EGR-1 transcription factors has provided a molecular paradigm for regulation of a coordinate set of target genes during cell growth and differentiation. The next critical questions that must be addressed are: what are the target genes which are normally regulated by WT1-EGR-1, and how does misregulation of these genes result in tumor formation?

ACKNOWLEDGMENTS

I thank the members of my laboratory who contributed to this work, Vikas Sukhatme for the CMV-Egr-1 and p3XEBS-CAT DNAs, and Tom Curran for many helpful discussions on growth-factor-induced genes. Many thanks to Maria Marinelli for preparing the manuscript. These studies were supported in part by grants CA52009, CA47983 to FJR and core grant CA10815 from the National Institutes of Health, and grants from the W. W. Smith Charitable Trust, the Hansen Memorial Foundation, and the Mary A. H. Rumsey Foundation. FJR is a Pew Scholar in the Biomedical Sciences. This paper is dedicated to the memory of Dr. Frank J. Rauscher, Jr.

REFERENCES

1. L.C. Cantley, K.R. Auger, C. Carpenter, B. Duckworth, A. Graziani, R. Kapeller and S. Soltoff, Oncogenes and signal transduction, *Cell* 64:281 (1991).
2. T. Hunter, Cooperation between oncogenes, *Cell* 64:249 (1991).
3. M. Cross and T.M. Dexter, Growth factors in development, transformation, and tumorigenesis, *Cell* 64:271 (1991).
4. R.L. Erikson, D. Alcorta, P.-A. Bedard, J. Blenis, H.-P. Biemann, E. Erikson, S.W. Jones, J.L. Maller, T.J. Martins and D.L. Simmons, Molecular analyses of gene products associated with the response of cells to mitogenic stimulation, *in*: "Cold Spring Harbor Symposia on Quantitative Biology Volume LIII." The Cold Spring Harbor Laboratory (1988).
5. V.P. Sukhatme, Early transcriptional events in cell growth: the Egr family, *J. Am. Soc. Nephrol.* 1:859 (1990).
6. V.P. Sukhatme, X. Cao, L.C. Chang, C.-H. Tsai-Morris, D. Stamenkovich, P.C.P. Ferreira, D.R. Cohen, S.A. Edwards, T.B. Shows, T. Curran, M.M. LeBeau and E.D. Adamson, A zinc-finger encoding

gene co-regulated with *c-fos* during growth and differentiation and after cellular depolarization, *Cell* 53:37 (1988).

7. D. Haber and D.E. Housman, The genetics of Wilms' tumor, *in*: "Advances in Cancer Research" 59:41 (1992).
8. A.G. Knudson and L.C. Strong, Mutation and cancer: a model for Wilms' tumor of the kidney, *J. Natl. Cancer Inst.* 48:313 (1972).
9. K.M. Call, T. Glaser, C.Y. Ito, A.J. Buckler, J. Pelletier, D.A. Haber, E.A. Rose, A. Kral, H. Yeger, W.H. Lewis, C. Jones and D.E. Housman, Isolation and characterization of a zinc-finger polypeptide gene at the human chromosome 11 Wilms' tumor locus, *Cell* 60:509 (1990).
10. M.Gessler, A. Poustka, W. Cavenee, R.L. Neve, S.H. Orkin and G.A.P. Bruns, Homozygous deletion in Wilms' tumors of a zinc-finger gene identified by chromosome jumping, *Nature* 343:774 (1990).
11. J. Pelletier, W. Bruening, C.E. Kashtan, S.M. Mauer, C.J. Manivel, J.E. Streigel, D.C. Houghton, C. Junien, R. Habib, L. Fouser, R.N. Fine, B.L. Silverman, D.A. Haber and D.E. Housman, Germline mutations in the Wilms' tumor suppressor gene are associated with abnormal urogenital development in Denys-Drash syndrome, *Cell* 67:437 (1991).
12. K. Pritchard-Jones, K. Fleming, D. Davidson, W. Bickmore, D. Porteous, C. Gosden, J. Bard, A. Buckler, J. Pelletier, D. Housman, V. van Heyningen and N. Hastie, The candidate Wilms' tumor gene is involved in genitourinary development, *Nature* 346:194 (1990).
13. F.J. Rauscher III, J.F. Morris, O.E. Tournay, D.M. Cook and T. Curran, Binding of the Wilms' tumor locus zinc finger protein to the EGR-1 consensus sequence, *Science* 250:1259 (1990).
14. J.F. Morris, S.L. Madden, O.E. Tournay, D.M. Cook, V.P. Sukhatme and F.J. Rauscher III, Characterization of the zinc finger protein encoded by the WT1 Wilms' tumor locus, *Oncogene* 6:2339 (1991).
15. N.P. Pavletich and C.O. Pabo, Zinc finger-DNA recognition: crystal structure of a Zif268-DNA complex at 2.1A, *Science* 252:809 (1991).
16. S.L. Madden, D.M. Cook, J.F. Morris, A. Gashler, V.P. Sukhatme and F.J. Rauscher III, Transcriptional repression mediated by the WT1 Wilms' tumor gene product, *Science* 253:1550 (1991).
17. I.A. Drummond, S.L. Madden, P. Rowher-Nutter, G.I. Bell, V.P. Sukhatme and F.J. Rauscher III, Repression of the insulin-like growth factor-II gene by the Wilms' tumor suppressor WT1, *Science* 257:674 (1992).
18. S.L. Madden, D.M. Cook and F.J. Rauscher III, A structure-function analysis of transcriptional repression mediated by the WT1, Wilms' tumor suppressor protein, *Oncogene* 8:1713 (1993).

ENZYME DEFICIENCY AND TUMOR SUPPRESSOR GENES: ABSENCE OF 5'-DEOXY-5'-METHYLTHIOADENOSINE PHOSPHORYLASE IN HUMAN TUMORS

Fulvio Della Ragione, Adriana Oliva, Rosanna Palumbo,
Gian Luigi Russo and Vincenzo Zappia

Institute of Biochemistry of Macromolecules
Medical School
Second University of Naples
Via Costantinopoli 16
80138, Naples, Italy

INTRODUCTION

Malignant transformation is due to mutations that modify the mechanisms regulating normal cellular growth and development. These alterations include the somatic activation of cancer-promoting genes[1] (cellular oncogenes) and the germline or somatic inactivation of tumor suppressor genes, also known as antioncogenes or recessive oncogenes[2]. While the identification of oncogenes is facilitated by their ability to transform appropriate host cells[3], the search for tumor suppressor genes is remarkably complicated by the lack of strong selection procedures.

However, detailed molecular genetic studies employing restriction fragment length polymorphism or polymerase chain reaction methods[4], along with extensive karyological analyses, have shown that non-random loss of some specific genetic material occurs in a large number of cancers and is often involved in the development or progression of the malignancy. This view is further strengthened by the observation of tumor suppression following monochromosome transfer in malignant cell lines[5].

Although these studies hint at the presence of numerous tumor suppressor genes, few recessive oncogenes have been isolated and cloned so far, including the retinoblastoma gene[6-8] (RB1 gene), the neurofibromatosis gene[9] (NF1 gene), three genes involved in the development of colorectal carcinoma[10-12] (p53, DCC and MCC genes), the gene whose loss of function is probably responsible for the Wilms' tumor[13,14] (WT1), and a gene frequently deleted in renal cell carcinoma and lung carcinoma[15] (PTP gene).

Among these antioncogenes the most thoroughly characterized is the RB1 gene whose cloning has been greatly facilitated by the observation that a large number of retinoblastoma cell lines are deficient in esterase D enzymatic activity. Indeed, starting from the cDNA of this

Advances in Nutrition and Cancer, Edited by
V. Zappia *et al.*, Plenum Press, New York, 1993

hydrolase, it has been possible to isolate an RB1 cDNA clone by means of the chromosome walking technique[7].

In this scenario, it is of interest the discovery that a high percent of malignant human cell lines[16] and of specimens from acute lymphoblastic leukemias and gliomas[17,18] are devoid of 5'-deoxy-5'-methylthioadenosine phosphorylase (MTAase) activity. Conversely, this enzymatic activity has been found in all normal tissues and cell lines of non-malignant origin investigated so far[17,19-22].

MTAase (5'-deoxy-5'-methylthioadenosine:orthophosphate methylthioribosyltransferase, EC 2.4.2.28) catalyzes the phosphorolytic cleavage of 5'-deoxy-5'-methylthioadenosine (MTA), a sulfur adenosyl nucleoside formed from S-adenosylmethionine by several independent pathways[23-25]. The reaction products, namely adenine and 5-methylthioribose 1-phosphate, are then recycled to AMP and to methionine, respectively[24,26]. Therefore, the enzyme presumably plays a key role in a purine salvage pathway and in the recycling of methylthio groups[25].

By means of mouse-human somatic cell hybridization studies, the putative gene for the phosphorylase has been mapped at the 9pter-9q12 region[27]. Since non-random abnormalities at 9p have been reported in several malignancies, it can be hypothesized that a chromosomal aberration encompassing the phosphorylase gene and some putative tumor suppressor gene might be responsible for the occurrence of the enzymatic defect exclusively in malignant cells. It should be underlined that the presence of a recessive oncogene on the short arm of chromosome 9 has also been suggested by chromosomal microcell transfer studies[5].

The aim of this paper is to review studies carried out in our laboratory on the purification and characterization of mammalian MTAase and on the possible presence of inactive form(s) of the phosphorylase in cell lines lacking this enzymatic activity. These results will be discussed in the light of the possible chromosomal linkage between the MTAase gene and a recessive oncogene. Moreover, we will discuss the Literature data that argue in favor of the occurrence of a putative tumor suppressor gene on the 9p region and that provide inference on its possible role in some specific cancers.

PURIFICATION AND CHARACTERIZATION OF MAMMALIAN 5'-DEOXY-5'-METHYLTHIOADENOSINE PHOSPHORYLASE

As stated in the Introduction, the importance of the purification and detailed characterization of mammalian MTAase has greatly increased since the discovery of a possible relationship between the deficiency of this enzymatic activity and malignancy. However, although many reports on this enzyme have been published[16-20], the mammalian protein has been purified to homogeneity and characterized only from human placenta[21] and bovine liver[22].

Table 1 reports the steps of the procedure developed in our laboratory for the purification of the bovine enzyme, that takes advantage of the remarkable stability of the enzyme against acid and thermal denaturation. The final specific activity from four different preparations ranged between 9 and 12 μmol of MTA cleaved per min per mg of protein. This represents more than 10,000-fold purification over the crude supernatant.

Approximately 2 mg of the homogeneous enzyme were obtained per kg of bovine liver, a quantity by far larger than that achievable from human placenta, and amply sufficient for a detailed physico-chemical characterization. Table 2 summarizes the main properties of the bovine protein. The molecular weight of the native enzyme calculated by analytical ultracentrifugation is 98,000 ± 3,000. A similar value has been estimated by gel filtration on Sephacryl S-200 using both the pure enzyme and the 20,000xg supernant, thus suggesting that no artifact occurred during the purification procedure (data not shown).

Table 1. Purification of MTAase from bovine liver.

Step	Total protein mg	Specific activity units*/mg	Yield %	Purification fold
Supernatant at 20,000xg	203,000	0.001	100	1
pH 5,2 treatment	119,000	0.0017	99	1.7
Heat treatment	66,000	0.0027	87	2.7
Ammonium sulphate (45-70%)	36,000	0.0034	66	3.4
Acetone (40-60%)	20,600	0.0060	61	6.0
DEAE-Sephacel	696	0.17	58	170
Chromatofocusing	21	2.1	21	2,100
Sephacryl S-200	2.4	10.3	12	10,300

*One unit is the amount of enzyme which cleaves 1 μmol of MTA per min at 37°C

The electrophoretic pattern of bovine liver enzyme under denaturing conditions showed no evidence of subunit heterogeneity. From plots of migration of marker protein versus their molecular weight, the average M_r of MTAase subunits was estimated to be 32,000 ± 500 (data not shown).

By means of circular dichroism the secondary structure of the protein was investigated; the α-helix and the β-structure contents were estimated to be approximately 5% and 48%, respectively, with about 47% random coil.

Altogether these studies indicate that, like the human enzyme[21], bovine MTAase is a trimer of about 98 kDa composed of apparently identical subunits. This protein structure, which is quite rare among enzymes, is shared by mammalian purine nucleoside phosphorylase (PNP). The resemblance between the two phosphorylases can also be observed regarding other hydrodynamic features such as frictional ratio, axial ratio, and Stokes radius, as well as the types of secondary structure. Indeed, a very low amount of α-helix and a relatively high percentage of random coil have been determined. However, as reported below, the lack of cross-reactivity between PNP and anti-MTAase immunoglobulins ruled out the occurrence of common epitopes and of notable sequence homologies.

In the following paragraph, we will report the studies carried out to clarify the mechanism(s) of the enzyme deficiency by antibodies against MTAase.

Table 2. Physico-chemical properties of bovine liver MTAase.

Property	Value
Molecular weight	
by gel filtration	96,000 ± 3,000
by sedimentation equilibrium	98,000 ± 3,000
Subunit molecular weight	32,000 ± 1,000
Sedimentation coefficient, $s_{20,w}$ (x10^{-13} cm/s)	6.11
Stokes radius (Å)	1.22
Frictional ratio	4.5
Axial ratio	1:4
Secondary structure	
% α-helix	5
% β-structure	48
% random coil	47

ABSENCE OF 5'-METHYLTHIOADENOSINE PHOSPHORYLASE IN HUMAN TUMORS AND MALIGNANT CELL LINES

Preliminary experiments demonstrated that native MTAase is scarcely immunogenic. Therefore, three different modified forms of the bovine protein, namely polymeric MTAase, hemocyanin-MTAase, and agarose-linked MTAase, were employed to prepare polyclonal antibodies[22]. Among the antisera obtained, two showed a remarkably high titer, i.e., one of those directed to hemocyanin-MTAase (AKMl) and the other raised against polymeric MTAase (APM). Moreover, an initial characterization demonstrated that these sera reacted with different epitopes of the protein, in that AKMl was able to recognize both the native and denatured form of the enzyme, while APM reacted only with the denatured phosphorylase[22].

Because of their different behavior, both antisera were employed for the search of inactive forms of MTAase in enzyme-deficient cells by means of immunoblotting and immunotitration methods. When AKMl was used to analyze 15,000 x g supernatant of several bovine tissues, only one band at 32 kDa was obtained, which perfectly corresponded to that of pure bovine MTAase standard (Fig. 1).

A densitometric analysis of bands, that allows a rough estimation of the protein content, gave results comparable to those calculated by enzymatic assay (data not reported). This result suggested that the phosphorylase was the only protein detected by the antiserum. However, since several structural features are shared by mammalian MTAase and PNP[28], cross-reactivity between anti-MTAase antibodies and PNP could be hypothesized.

As shown in Fig. 2, SDS-polyacrylamide gel electrophoretic analysis of pure bovine liver MTAase (lane 3) and a highly purified preparation of bovine spleen PNP (lane l) showed an identical subunit molecular weight. In contrast, no cross-reactivity of the spleen phosphorylase towards APM or AKMl (data not reported) was observable by dot-blot analysis.

AKMl serum, which also reacts with native MTAase, was used to set up a standard curve for quantitation of the protein by immunotitration. In particular, increasing amounts of the pure bovine or human phosphorylase were immunoprecipitated with different quantities of immunoglobulins and the results plotted as the amount of IgG needed to obtain a 50% loss of

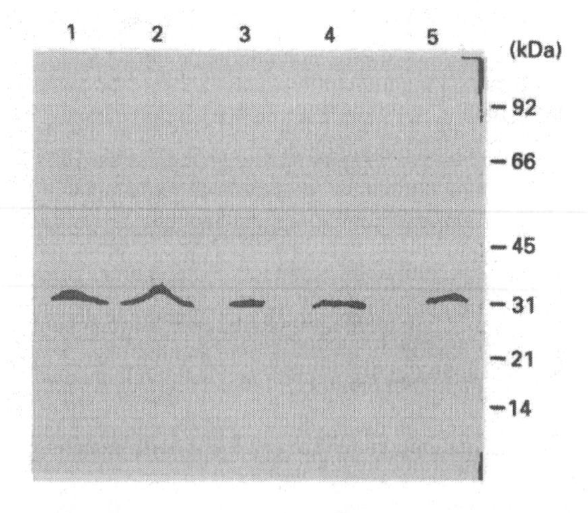

Figure 1. Immunoblotting analysis of various bovlne tissues extracts by AKM1 serum.
Lane 1, liver; lane 2, spleen; lane 3, pure liver enzyme; lane 4, testis; lane 5, brain. Each lane contains about 100 μg of 15,000xg supernatant, except lane 3, which contains 50 ng of homogeneous MTAase. Samples were run on a 12.5%-polyacrylamide gel under denaturing conditions, transferred on a nitrocellulose paper and then processed as in Ref. 22. Reprinted with permission from Ref. 22.

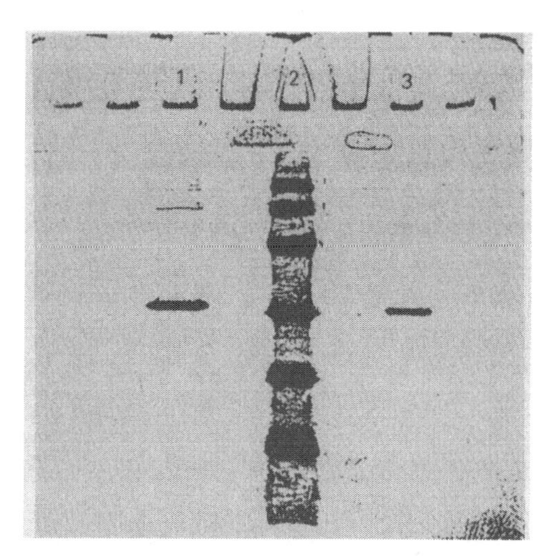

Figure 2. SDS-Gel electrophoretic analysis of highly purified bovine spleen purine nucleoside phosphorylase and pure bovine liver MTAase.
Lane 1, 5 μg of purine nucleoside phosphorylase; lane 2, molecular weight standards (from top to bottom: 94 kDa, 67 kDa, 43 kDa, 30 kDa, 20 kDa, 14 kDa); lane 3, 3 μg of pure MTAase. Reprinted with permission from Ref. 22.

activity (Fig. 3). The standard curve generated was linear in the range of 0.5-10 ng of MTAase with a slope indicating the immunoprecipitation of 1 ng of human enzyme by 6 μg of IgG.

Different mechanisms can explain the absence of MTAase activity in human malignant cell lines. First, the cells might contain some inhibitor interfering with the enzymatic activity. Second, the protein might not be synthesized as a consequence of a genetic alteration hampering gene transcription or resulting in the production of untranslatable or unstable mRNA. Finally, the deficient cells might synthesize an inactive form of the protein. In the latter case, immunological analyses should ascertain the occurrence of cross-reacting material (CRM) in the negative cell lines.

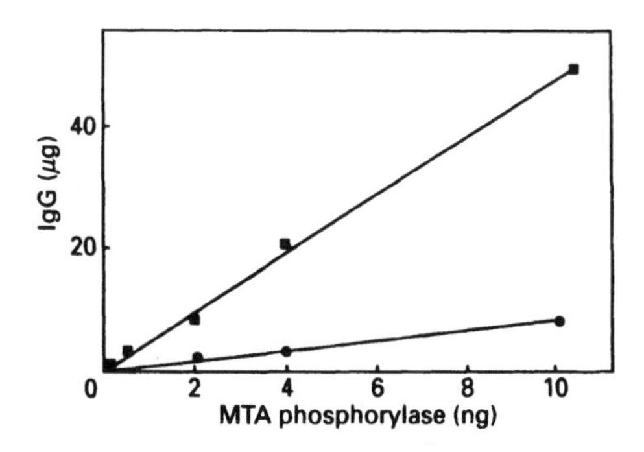

Figure 3. Immunoprecipitation standard curve for MTAase estimation.
Different amounts of human placenta (■) and bovine liver (●) MTAase were immunoprecipitated by different quantities of AKMI. The amount of IgG which gave 50% of precipitation was then plotted against MTAase amount. The results are the average of three different estimations Reprinted with permission from Ref. 22.

Table 3. Effect of MTAase deficient cell-line extracts on the phosphorylase activity.

Phosphorylase source	Cell extracts added	Enzyme activity (units x10^6)
Human placenta*	None	10.7
K562**	None	0
Jurkat**	None	0
Human placenta	K562	12
Human placenta	Jurkat	11
HeLa°	None	30
HeLa	K562 + Jurkat	32
Human placenta°°	K562 + Jurkat	8.3
HeLa°°	None	28
HeLa°°	K562 + Jurkat	29

*1 ng of human placental enzyme; ** 500 μg protein of K562 or Jurkat cell extracts; °25 μg protein of HeLa cell extract; °° samples preincubated for 2 hrs at 37°C.

As shown in Table 3, the addition of high amounts of two MTAase-deficient cell lines extracts (K562 and Jurkat) to the pure protein or to a MTAase-containing cell homogenate (HeLa) did not modify the enzymatic activity. This was also observed when the various mixtures were preincubated at 37°C for 2 hrs. These results permitted the exclusion of both a possible enzymatic inhibitor and highly active proteases in the enzyme-deficient cells.

When various MTAase-positive human cells were analyzed by immunoblotting, a band at 30-32 kDa was clearly evidentiable. Figure 4 reports a Western blot analysis of normal and

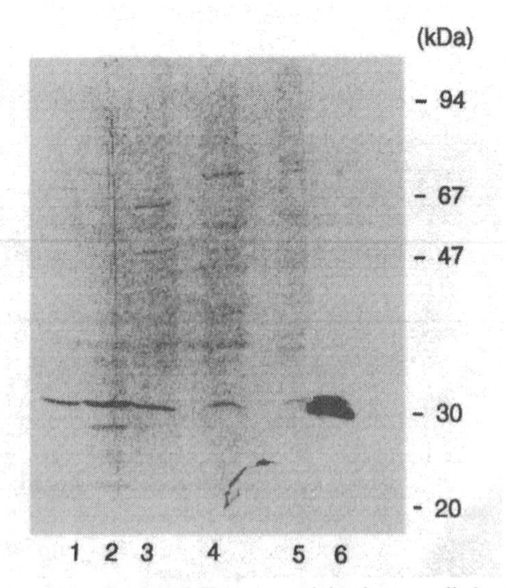

Figure 4. Immunoblotting analysis of various MTAase-containing human cells by AKMI serum. Lane 1, A431; lane 2, HeLa: lane 3, human fibroblasts, lanes 4 and 5, human leukemic samples, lane 6, human placenta pure enzyme. Each lane contains about 500 μg of 15,000xg supernatant except lane 6, which contains 200 ng of the homogeneous protein. On the right side, masses of protein markers are indicated in kDa. The experiments were carried out as described in Fig.l. Reprinted with permission from Ref. 22.

Table 4. MTAase content in cell-line extracts determined by enzymatic assays and immunotitration method.

Source	10^{-4} x MTAase activity (units/mg)	MTAase content (ng/mg)
HeLa	11	103
Human fibroblasts	7	60
A431	18	173
KB	15	125
Jurkat	-	< 0.5
K562	-	< 0.5

MTAase activity was determined by enzymatic assay; MTAase content was calculated by immunotitration method.

malignant cells, i.e. HeLa cells (lane 1), A431 (lane 2), human fibroblasts (lane 3), two samples of leukemic cells (lanes 4 and 5), and a phosphorylase standard (lane 6).

An immunotitration analysis was also carried out on the same samples employing AKMl serum to obtain an estimation of the MTAase content (Table 4). The results are in good agreement with those of immunoblot. Moreover, Table 4 shows that no CRM was demonstrable in K562 and Jurkat cells at least within the sensitivity range of the method.

Figure 5 shows an immunoblotting analysis of two enzyme-deficient cell lines and of the HeLa cells. While a clear band was detected in the positive cells (lane 3), no band was observed in the same area in K562 and Jurkat cell lines (lanes 1 and 2, respectively). In Jurkat cells a strong band was seen at higher molecular weight. However, the same band was detected with a control serum and therefore is not related to possible CRM (data not reported).

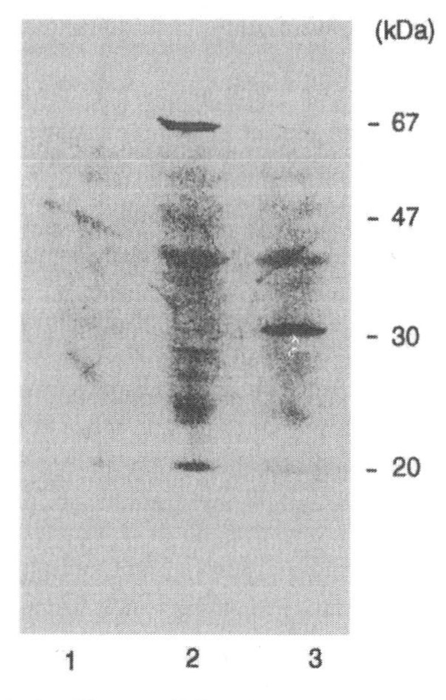

Figure 5. Immunoblotting analysis of human cell lines by AKM1 serum.
Lane 1, K562; lane 2, Jurkat; lane 3, HeLa. Each lane contains 500 μg of 15,000xg supernatant. On the right side, masses of protein markers are indicated in kDa. The experiments were carried out as in Fig.1. Reprinted with permission from Ref. 22.

The absence of the phosphorylase protein in the cells deficient in this enzymatic activity might be explained by several alterations involving any step of the protein synthetic process, namely: i) a genetic aberration that leads to a lack of MTAase gene expression; ii) synthesis of an unfunctional mRNA; or iii) synthesis of an unstable protein. In order to evaluate the latter possibility, the biosynthesis of MTAase was investigated in phosphorylase-positive (HeLa) and negative cells (K562 and Jurkat) by pulse-labeling experiments with (^{35}S)methionine. To select the labeling-time, preliminary experiments on the half-life of the protein were carried out by employing actinomycin D and cycloeximide as inhibitors of protein synthesis. The extrapolated value in phosphorylase-containing cells, i.e. HeLa, A431 and human fibroblasts, was higher than 48 hrs (data not reported). Therefore, due to the long turnover, a labeling of 4 hrs was used.

HeLa and Jurkat cells were incubated with (^{35}S)methionine and the cell extracts precipitated with AKMI serum. The immunoprecipitated material was then analyzed by SDS-polyacrylamide gel electrophoresis and visualized by autoradiography after a prolonged exposition. As shown in Fig. 6, a specific band was observed at 30 kDa (lanes 2 and 3) in the HeLa extracts. Conversely, no band was observed in Jurkat extracts at this molecular weight (lanes 5 and 6). Lanes 1 and 4 contain HeLa and Jurkat extracts immunoprecipitated with control sera. An identical result was obtained for K562 cell line (data not reported). In additional experiments, subconfluent cultures of HeLa and Jurkat cells were grown for 2 hrs in the presence of (^{32}P)orthophosphate, immunoprecipitated with AKMI and analyzed by SDS-polyacrylamide gel electrophoresis. No labeling was observed in the 30-32 kDa area, thus indicating the absence of post-synthetic phosphorylation of the protein (data not shown).

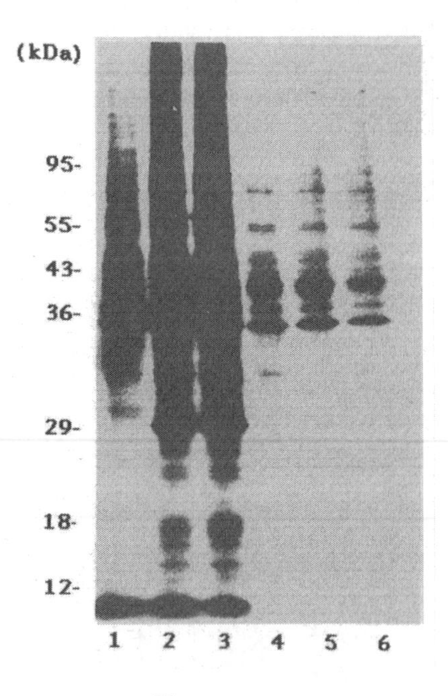

Figure 6. Immunoprecipitation analysis of (^{35}S)-labeled extracts from MTAase positive and negative cells.

Lanes 1-3 contain HeLa labeled cell extracts; lanes 4-6 contain Jurkat labeled cell extracts. Lanes 1 and 4 were immunoprecipitated with a control serum while lanes 2, 3, 5, and 6 with AKMI serum. Each immunoprecipitate was run on a gradient (7.5-15%) polyacrylamide gel under denaturing conditions. The gel was then dried and analyzed as described in the Experimental Procedures. On the left side, masses of protein markers are indicated in kDa. Reprinted with permission from Ref. 22.

These results clearly demonstrated that the absence of MTAase activity is due to the lack of the protein itself and not to the occurrence of inactive or unstable forms of the enzyme. Considerable evidence hints at this conclusion. Indeed, the cell lines deficient in the phosphorylase activity did not contain any inhibitor or protease that could inactivate endogenous MTAase (Table 1). No MTAase protein was detected by Western blotting analysis employing two highly specific antisera raised against different immunogenic forms of the protein (Fig. 5), and an identical result was obtained by using a sensitive immunotitration method (Table 3). No labeled protein was observed when the biosynthesis of MTAase was investigated by pulse-labeling experiments with (^{35}S)methionine (Fig. 6).

Two different mechanisms might explain the absence of phosphorylase protein in K562 and Jurkat cells: (i) the negative cells present an alteration of MTAase gene which prevents the phosphorylase synthesis; or (ii) the deficiency is due to some epigenetic mechanism(s) that hamper(s) the expression of a normal MTAase gene. It could be hypothesized that the malignant deficient cells derive from a normal counterpart that is in a particular stage of differentiation lacking MTAase protein. Although the latter hypothesis deserves further investigation, it appears quite unlikely since MTAase must be envisioned as a housekeeping enzyme. Indeed, it shows a long turnover rate[29] and is scarcely inducible[24]. Therefore, the presence of primary genetic alteration(s) appears to be the most probable cause of the enzyme deficiency.

It is important to stress that the same conclusion has been reached by a successive study[30] carried out investigating the deficiency of MTAase in human glioma cell lines and specimens from primary glioblastomas or astrocytomas. Even in these cases the absence of enzymatic activity strictly corresponds to the absence of protein.

Since the absence of the enzyme protein is complete (partial deficiency has never been demonstrated), it is conceivable that the genetic alteration of the MTAase gene must occur in a homozygous condition. This conclusion, along with the high incidence of the enzymatic deficiency, strongly suggests that the phosphorylase gene maps quite near to some tumor suppressor gene. This hypothesis is strongly strengthened by data reported in the Literature and discussed in the next paragraph.

9p21-22 REGION ABERRATIONS IN HUMAN TUMORS AND PUTATIVE TUMOR SUPPRESSOR GENE(S)

The molecular evidence for the existence of tumor suppressor genes is largely circumstantial. The initial indications of a recessive oncogene locus came generally from cytogenetic studies documenting non-random chromosome alterations, including deletions, seen in metaphase spreads of various cancers. More important and intellectually satisfying data, were obtained from molecular analyses, utilizing the technique of restriction fragment length polymorphism analysis. It has to be underlined that the chromosome regions analysed by specific probes were usually selected on the basis of previous cytogenetic studies.

Cytogenetic abnormalities at the 9p region have been described in several tumors including acute lymphoblastic leukemia[31] (ALL), non-Hodgkin's lymphoma[32], melanoma[33], malignant glioma[34], and non-small cell lung carcinomas[35] (NSCLC).

In the hematopoietic tumors, including ALL and non-Hodgkin's T-lymphomas, deletions or unbalanced translocations of the short arm of chromosome 9 have been reported with frequencies of 7 to 13 percent [31,36,37]. Some authors[37] reported a very high incidence (80%) in patients affected by ALL with a clinical syndrome characterized by bulky diseases in sites other than bone marrow, particularly within lymph nodes, spleen, and mediastinum, and by high white-cell counts in association with blasts bearing T-cell receptors, the so-called "lymphomatous ALL". However, additional studies did have not yet confirmed these results.

Approximately 50% of human gliomas have structural rearrangements affecting chromosome 9[38] and the incidence of 9p alterations in malignant melanoma are about 46%[39,40].

Finally, Lukeis et al.[35] reported a cytogenetic analysis of ten primary non-small cell lung carcinomas, including five adenocarcinomas, three squamous cell, and two large cell carcinomas, which was carried out in an attempt to determine karyotypic changes involved in the early stage of disease. The tumors studied were aneuploid and exhibited complex karyotypes with multiple structural and numerical abnormalities. Clonal structural rearrangements were identified, and in particular loss of material from the short arm of chromosome 9 had a 90% incidence. This loss was due to non-reciprocal translocation, deletion, or chromosome loss. The breakpoints were in the region 9q13 to p22. The authors concluded that, while a primary cytogenetic change in NSCLC has not been conclusively identified, their findings implicate loss of material from 9p as a potentially important event.

Molecular biology analyses of chromosome 9p have been carried out by both classical restriction fragment length polymorphism studies and linkage analyses employing polymerase chain reaction amplification of selected regions. The main probes of the region include cDNA of the α-interferon (IFNA) gene cluster, the β-interferon (IFNB1) locus, the glycoprotein 4β-galactosyltransferase (GGTB2) gene, the tyrosinase-related protein (TYRP) gene, and the argininsuccinate synthase pseudogene 3 (ASSP3). Additional markers of the 9p chromosome are D9S33, D9S126, D9S3 and D9S19.

A quite detailed genetic and physical map of the region surrounding the interferon genes on 9p has recently been created and the suggested order is D9S33, IFNB1, IFNA, D9S126, D9S3, D9S19, GGTB2 and ASSP3[41]. Independent studies, carried out on human glioma cell lines and primary gliomas, in order to finely map the interferon cluster genes, allowed the conclusion that IFNB1 is the telomeric gene while IFNA8, IFNWW2, IFNWP19 and IFNA1 are the centromeric genes. By using the mentioned probes and by taking into consideration the order of the genes on the 9p chromosome, several groups have investigated at the molecular level the genetic alterations of the area under study in different tumors in order to characterize the region that should contain the putative tumor suppressor gene.

Structural rearrangements of 9p21-22 in acute leukemia employing IFNA and IFNB1 probes are around 30%[42]. By using the same markers, 67% of glioma-derived cell lines had hemizygous or homozygous deletions or rearrangement of sequences around interferon genes, while 37% of primary glioma tumor samples had deletions of these genes[43]. The incidence increased to 50-60%[43,44] if only high grade malignant gliomas were considered. Finally, very recent results from malignant melanomas indicate that 86% of tumor specimens or cell lines derived from this cancer contain hemizygous or homozygous aberration at 9p[40]. These results are strongly indicative (and probably conclusive) of the occurrence of a tumor suppressor gene in the 9p21-22 area.

As far as the most informative probes are concerned, it is highly probable that the tumor suppressor gene maps between the centromeric side of IFNA cluster (IFNA8, IFNWW2, IFNWP19 and IFNA1 genes) and the D9S126 marker. This region of 2-3 million bases should contain the putative recessive oncogene.

Since the MTAase gene has not yet been cloned, no data regarding its precise localization and its relationship with the above mentioned genes or 9p markers are available

However, the data on the cell lines and tumors lacking the phosphorylase activity might represent indirect information on the homozygous loss of function of MTAase gene. When the alteration of IFNA and IFNB1 was studied contemporaneously with he lack of MTAase activity, the absence of the phosphorylase appeared to occur more frequently than the homozygous aberration of IFNA or IFNB1. Therefore the MTAase gene appears to be nearer than interferon genes to the tumor suppressor gene. Additional findings indicate that the IFNA gene maps at the telomeric end of the region studied while the MTAase gene at the centromeric extremity. Therefore, given the localization of D9S126 marker (that also maps at the centromeric end of the 9p region containing the putative tumor suppressor gene), studies might be carried out by comparing the homozygous alteration of D9S126 and the absence of MTAase activity to better establish the relationship between these two probes.

Finally, as for the role of this tumor suppressor gene, some indirect considerations might be made. First, since the incidence of ALL and possibly non-Hodgkin's lymphomas that show alteration at 9p21-22 is around 30% (at least only with indirect probes like IFNA), the role of this putative recessive oncogene is related to the development of tumors. However, the hematopoietic tumors analyzed are quite heterogenous and a conclusive hypothesis might expect more detailed investigations. Second, the incidence of aberration at the analyzed region in malignant gliomas and melanomas is very high, especially considering that no direct probes are employed. Apparently, the role of this putative antioncogene is different in these two types of tumors, since in malignant melanoma the functional loss is probably the initial event in the development of cancer, while in the case of the gliomas it is involved in the progression process. However, we must underline the common embryological origin of melanocytes and glioma cells from the neuronal crests. So in the neuroectodermal tissue this tumor suppressor gene could play a pivotal role. Finally, the data from Lukeis et al.[35] on the NSCLC are very impressive especially in view of the high incidence of lung cancer. Future studies will clarify the role and meaning of this new putative tumor suppressor gene.

REFERENCES

1. J.M. Bishop, The molecular genetics of cancer, *Science* 235:305 (1987)
2. R.A. Weinberg, Oncogenes, antioncogenes, and the molecular bases of multistep carcinogenesis, *Cancer Res*. 49:3713 (1989)
3. E.P. Reddy, A.M. Skalka, T. Curran, eds. (1988) The Oncogene Handbook, Elsevier Science Publisher, Amsterdam
4. B.Ponder, Gene losses in human tumours, *Science* 335:400 (1988)
5. R. Sager, Tumor suppressor genes: the puzzle and the promise, *Science* 246:1406 (1989)
6. S.H. Friend, R. Bernards, S. Rogeij, R.A. Weinberg, J. M.Rapaport, D.M. Albert, and T.P. Dryia, A human DNA segment with properties of the gene that predisposes to retinoblastoma, Nature 323:643 (1986)
7. W.H. Lee, R. Bookstein, F. Hong, L.-J. Young, J.-Y. Shew, E.Y.-H.P. Lee, Human retinoblastoma susceptibility gene:cloning identification and sequence, *Science* 235:1394 (1987)
8. Y.-K.T. Fung, A.L. Murphree, A. T'Ang, J. Qian, S.H. Hinrichs, and W.F. Benedict, Structural evidence for the authenticity of the human retinoblastoma gene, *Science* 236:1657 (1987)
9. D. Viskohil, A. Buchberg, G. Xu, R. Cawthon, J. Stevens, R. Wolff, M. Culver, J. Carey, N. Copeland, N. Jenkins, R. White, and P. O'Connel, Deletions and a translocation interrupt a cloned gene at the neurofibromatosis type 1 locus, Cell 62: 187 (1990)
10. S.J. Baker, E.R. Fearon, J.M. Nigro, S.R. Hamilton, A.C. Preisinger, J.M. Jessup, P.Van Tuinen, D.H. Ledbetter, D.F. Barker, Y. Nakamura, R. White, B. Vogelstein, Chromosome 17 deletions and p53 gene mutations in colorectal carcinomas, *Science* 244:217 (1989)
11. E.R. Fearon, K.R. Cho, J.M. Nigro, S.E. Kern, J.W. Simons, J.M. Ruppert, S.R. Hamilton, A.C. Preisinger, G. Thomas, K.W. Kinzler, B. Vogelstein, Identification of a chromosome 18q gene that is altered in colorectal cancers, *Science* 247:49 (1990)
12. K.W. Kinzler, M.C. Nilbert, L.K. Su, B. Vogelstein, T.M. Bryen, D.B. Levy, K.J. Smith, A.C. Preisinger, P. hedge, D. McKechinc, Identification of FAP locus gene from chromosome 5q21, *Science* 253:661 (1991)
13. E.A. Rose, T. Glaser, C. Jones, L.S. Cassandra, W.H. Lewis, K.M. Call, M. Minden, E. Champagne, L. Bonetta, H. Yeger and D.E. Housman, Complete physical map of the WAGR region of 11p13 localizes a candidate Wilms' tumor gene, *Cell* 60:495(1990)

14. K.M. Call, T. Glaser, C.Y. Ito, A.J. Buckler, J. Pelletier, D.A. Haber, E.A.Rose, A., Kral, H. Yeger, W.H. Lewis, C. Jones, and D.E. Housman, Isolation and characterization of a zinc-finger polypeptide gene at the human chromosome 11 Wilms' tumor locus *Cell*, 60:509 (1990)
15. S. LaForgia, B. Morse, J. Levy, G. Barnea, L.A. Cannizzaro, F. Li, P.C. Nowell, L. Boghosian-Sell, J. Glick, A. Weston, C.C. Harris, H. Drabkin, D. Patterson, C.M. Croce, J. Schlessinger, and K. Huebner, Receptor protein-tyrosine phosphatase γ is a candidate tumor suppressor gene at human chromosome region 3p21, Proc. Natl. Acad. Sci. USA, 88:5036 (1991)
16. N. Kamatani, W.A. Nelson-Rees and D.A. Carson, Selective killing of human malignant cell lines deficient in methylthioadenosine phosphorylase, a purine metabolizing enzyme, *Proc. Natl. Acad. Sci. USA* 78:1219 (1981)
17. N. Kamatani, A.L. Yu and D.A. Carson, Deficiency of methylthioadenosine phosphorylase in human leukemic cells in vivo, *Blood*, 60:1387 (1982)
18. S.T. Traweek, M.K. Riscoe, A.J. Ferro, R.M. Braziel, R.E. Magenis and J.H. Fichten, Methythioadenosine phosphorylase deficiency in acute leukemia : pathologic, cytogenetic, and clinical features, *Blood*, 71:1568 (1988)
19. J.J. Toohey, Methylthio group cleavage from methylthioadenosine. Description of an enzyme and its relationship to the methylthio requirement of certain cells in culture, *Biochem. Biophys: Res. Commun.*, 78:1273 (1977)
20. V. Zappia, A. Oliva, G. Cacciapuoti, P. Galletti, G. Mignucci and, M. Cartenì-Farina, Substrate specificity of 5'-Methylthioadenosine phosphorylase from human prostate, *Biochem. J.* 175:1043 (1978)
21. F. Della Ragione, M. Cartenì-Farina, V. Gragnaniello, M.I. Schettino and, V. Zappia, Purification and characterization of 5'-deoxy-5'-methylthioadenosine phosphorylase from human placenta J. *Biol. Chem.* 261:12324 (1986)
22. F. Della Ragione, A. Oliva, V. Gragnaniello, G.L. Russo, R. Palumbo and V. Zappia, Physicochemical and immunological studies on mammalian 5'-deoxy-5'-methylthioadenosine phosphorylase, *J. Biol. Chem.* 265:6241 (1990)
23. V. Zappia, M. Cartenì-Farina, G. Cacciapuoti, A. Oliva and, A.Gambacorta. Recent studies on the metabolism of 5'-methylthioadenosine (1980) in Natural Sulfur Compounds, Novel Biochemical and Structural Aspects (Cavallini, D., Gaull, G., & Zappia, V. eds.) pp. 133-148, Plenum Publishing Corp. New York
24. H.G. Williams-Ashman, J. Seidenfeld, P. Galletti, Trends in the biochemical pharmacology of 5'-deoxy-5'-methylthioadenosine, *Biochem. Pharmacol.*, 31:277 (1982)
25. F. Della Ragione, M. Cartenì-Farina and, V. Zappia, 5'-deoxy-5'-methylthioadenosine: novel metabolic and physiological aspects (1989) in The Physiology of Polyamines (Bachrach U. & Heimer Y.M. eds.) pp. 231-254, CRC Pregg, Inc., Boca Raton, Florida
26. P.S. Backlund Jr, C.P. Chang.and R.A Smith, Identification of 2-keto-4-methylthiobutyrate as an intermediate compound in methionine synthesis from 5'-methylthioadenosine, *J. Biol. Chem.* 257:4196 (1982)
27. C.J. Carrera, R.L. Eddy, T.B. Shows and D.A. Carson, Assignment of the gene for methylthioadenosine phosphorylase to human chromosome 9 by mouse-human somatic cells hybridization, *Proc. Natl. Acad. Sci. USA*, 81:2665 (1984)
28. J.D. Stoeckler, R.P. Agarwal, K.C. Agarwal, K. Schimd, and R.E. Parks Jr., Purine nucleoside phosphorylase from human erythrocytes: physicochemical properties of the crystalline enzyme, *Biochemistry* 17:278 (1978)
29. J. Seidenfeld, J. Wilson, J. and, H.G. Williams-Ashman, Androgenic regulation of 5'-deoxy-5'-methylthioadenosine concentrations and methylthioadenosine phosphorylase activity in relation to polyamine metabolism of rat prostate, *Bioch. Biophys. Res. Commun.* 95:1861 (1980)
30. T. Nobori, J.G. Karras, F. Della Ragione, T.A. Waltz, P.P. Chen and D.A. Carson, Absence of methylthioadenosine phosphorylase in human gliomas, *Cancer Res.*, 51:3193 (1991)

31. S.B. Murphy, S.C. Raimond, G.K. Rivera, M. Crone, R.K. Dodge, F.G. Behm, C.-H. Pui and, D.L. Williams, Non random abnormalities of chromosome 9p in childhood acute lymphoblastic leukemis: association with high-risk clinical features, Blood 74:409 (1989)
32. M.O. Diaz, S. Ziemin, M.M. Le Beau, P. Pitha, S.D. Smith, R.R. Chilcote and, J.D. Rowley, Homozygous deletion of alpha-a and beta-1 interferon genes in human leukemia and derived cell lines, *Proc. Natl. Acad. Sci. USA* 85:5259 (1988)
33. J.M. Cowan, R. Halaban, U. Francke, Cytogenetic analysis of melanocytes from premalignant nevi and melanomas, *J. Natl. Canc. Inst.* 80:1159 (1986)
34. S.H. Bigner, J. Mark, R.C. Burger, M.S. Mahaley Jr.,D.E. Bullard, L.H. Muhlbaier, and D.D. Bigner, Specific chromosomal abnormalities in malignant human gliomas *Cancer Res.* 48:405 (1988)
35. R. Lukeis, L. Irving, M. Garson, S. Hasthorpe , Cytogenetics of non-small cell lung cancer: analysis of consistent non-random abnormalities, *Genes-Chromosom-Cancer*; 2: 116 (1990)
36. J. Kowalcizyk, A.A. Sandberg, A possible subgroup of ALL with 9p-, *Cancer Genet. Cytogenet.* 9:383 (1983)
37. R.R. Chilcote, E.Brown and J.D. Rowley, Lymphoblastic leukemia with lymphomatous features associated with abnormalities of the short arm of chromosome 9, New Engl. J. Med.313:286 (1985)
38. S.H. Bigner, J. Mark, D.E. Bullard, S.M. Mahaley Jr., and D. Bigner, Chromosomal evolution in malignant human gliomas starts with specific and usually numerical deviations, *Cancer Genet. Cytogenet.* 22:121 (1986)
39. L.A. Cannon-Albright, D.E. Goldgar, L.J. Meyer, C.M. Lewis, D.E. Anderson, J.W. Fountain, M.E. Hegi, R.W Wiseman, E.M. Petty, A.E. Bale, O.I. Olopade, M.O. Diaz, D.J. Kwiatkowski, M.P. piepkorn, J.J. Zone, M.H. Skolnick, Assignment of a locus, for familial melanoma, MLM, to chromosome 9p13-p22, *Science* 255: 1148 (1992)
40. J.W. Fountain, M. Karayiorgou, M.S. Ernstoff, J.M. Kirkwood, D.R. Vlock, L. Titus-Ernstoff, B. Bouchard, S. Vijayasaradhi, A.N. Houghton, J. Lahti, V.J. Kidd, D.E. Housman, and N.C. Dracopoli, Homozygous deletions within chromosome band 9p21 in melanoma, *Proc. Natl. Acad. Sci. U.S.A.* 89: 10557 (1992)
41. J.W. Fountain, M. Karayiorgou, D. Taruscio, S.L. Graw, A.J. Buckler, D.C. Ward, N.C. Dracopoli, and D.E. Housman, Genetic and physical of the interferon region on chromosome 9p, Genomics, 14:105 (1992)
42. M.O Diaz, C.M. Rubin,A. Harden, S. Ziemin,. R.A. Larson, M.M. Le Beau and J.D. Rowely, Deletions of interferon genes in acute lymphoblastic leukemia, *New Engl. J. Med.* 322:77 (1990)
43. C.D. James, J. He, E. Carlbom, M. Nordenskjold, W.K. Cavenee, V.P. Collins Chromosome 9 deletion mapping reveals interferon alpha and interferon beta-1 gene deletions in human glial tumors,.*Cancer Res.* 51:1684 (1991)
44. O.I. Olopade, R.B. Jenkins, D.T. Ransom, K. Malik, H. Pomykala, T. Nobori, J.N. Cowan, J.D. Rowley, and M.O. Diaz, Molecular analysis of deletions of the short arm of chromosome 9 in human gliomas, *Cancer Res.* 52:2523 (1992)

HOX GENE EXPRESSION IN HUMAN CANCERS

Pasquale Barba, Maria Cristina Magli, Claudia Tiberio, and Clemente Cillo

International Institute of Genetics and Biophysics
Via Marconi 12
80125, Naples, Italy

Cancer is a complex, multistep process involving the aberrant expression of genes that regulate cell growth. The biochemical function of proto-oncogenes allows the pathway of cell growth to be divided into several important steps (1). One of these entails the trans-activation of target genes involved in cell proliferation (2). The products of several nuclear oncogenes (i.e. fos, jun, erb A)(3-5) as well as the tumor suppressor gene Rb (6), are DNA-binding proteins with transactivating activity. Thus, the role played by proteins that can exert effects on gene expression is becoming increasingly important to understanding the molecular basis of neoplasia.

Homeobox genes are a family of genes containing a common 183-nucleotide sequence. The homeobox encodes a 61 aminoacid domain, the homeodomain (HD), which includes a helix-turn-helix motif responsible for the DNA binding ability of homeobox-containing proteins (7). On the basis of structural similarities and direct evidence that *Drosophila* homeodomain proteins are capable of binding DNA sequences and modulating transcriptional activity, it is generally accepted that homeodomain proteins are transcriptional regulators (8). The homeobox was originally discovered in genes controlling *Drosophila* development (9) and has subsequently been isolated in other, evolutionarily distant species, such as nematodes and vertebrates (10). Different homeobox gene families have evolved which encode homeodomains of different types or classes. Among these HDs the *Drosophila Antennapaedia* (Antp) homeodomain defines one consensus sequence referred to as class I HDs (7). Mammalian class I homeobox genes are clustered in restricted regions of the genome (HOX loci) on four distinct chromosomes that presumably evolved by duplication of a primordial gene cluster (11). A striking finding is that the order of genes within each cluster is also highly conserved throughout evolution, suggesting that the physical organization of HOX genes may be essential for their expression (12). HOX genes are expressed during embryogenesis in a tissue-specific and frequently stage-related fashion (13). Expression of individual HOX genes has been detected in normal adult tissues (11, 17).

An association between genes that control transcription and the oncogenic process has recently been strengthened based on a number of independent observations. First, constitutive expression of the HOX 2.4 gene can be oncogenic in mice (14). Second, the pbx

Advances in Nutrition and Cancer, Edited by
V. Zappia *et al.*, Plenum Press, New York, 1993

homeobox gene in the t(1; 19) translocation of pre-B acute leukemias is aberrantly expressed (15). Third, altered expression of the tcl-3 (HOX11) gene in the t(10; 14) translocation has been reported in some T cell leukemias (16). Fourth, the coordinate regulation of HOX genes may play an important role in human hemopoietic differentiation (18).

Given this association between HOX genes, developmental processes and oncogenesis, we wished to determine whether the physical organization of HOX genes reflects a regulatory network involved in normal organogenesis and/or in neoplastic transformation. We have analyzed the expression of a panel of 38 HOX genes in normal adult human tissues or organs and in their neoplastic counterparts. Some of the HOX genes tested show marked differences in expression in cancer specimens when compared to normal organs, suggesting an association between altered HOX gene expression and cancer.

HOX GENE EXPRESSION IN NORMAL AND NEOPLASTIC HUMAN KIDNEY

The vertebrate kidney is an ideal model system for studying organogenesis, cell differentiation and neoplastic transformation (19). Transitory and vestigial pronephric and mesonephric structures are generated during early development, leading finally to the formation of the adult kidney, the methanephros. The development of the methanephric kidney component, the branching epithelium of the ureter and the mesenchyme converted into epithelial elements, occurs in a synchronous manner that is strictly controlled, both temporally and spatially.

Class I homeobox genes within the four HOX loci can be aligned horizontally, according to their physical position on each chromosome, and vertically on the basis of the maximal sequence homology of their homeodomains. This alignment defines 13 paralogous groups (20). Only two of them, namely groups 5 and 10, contain four HOX genes in all four loci (Figure 1).

We analyzed the expression of the four HOX gene clusters in normal human kidney and in tumor samples derived from patients with renal carcinomas. Poly (A)+ RNAs from normal kidney and from cancer specimens were hybridized by Northern blotting with probes containing the 3' untranslated regions specific for each of the 38 HOX genes (20-22), organized in four large clusters, HOX1-HOX4 located on chromosomes 7, 17, 12, and 2, respectively (see Figure 1).

The pattern of HOX gene expression in normal human adult kidney is shown in Figure 1. Of the 38 HOX genes tested, 30 are actively expressed. In normal kidney, HOX genes are switched on or off in blocks containing a variable number of contiguous genes within each locus. Expression of the genes at the extreme 5' and 3' ends of the HOX-1 locus, HOX-1J and HOX-1F, is undetectable, while 9 contiguous genes of the same locus, from HOX-1I to HOX-1K are actively expressed. Furthermore the entire HOX-2 locus is expressed, except for the 3' most gene HOX-2I. Moreover 3 contiguous genes, HOX-3G, HOX-3F, HOX-3H, at the 5' end of the HOX-3 locus , are silent, while a block of 6 genes (from HOX-3I to HOX-3E) are actively expressed. Finally, the two genes at the extreme 5' end of the HOX-4 locus , HOX-4I and HOX-4H, are silent and 7 contiguous genes, at the 3' end, are all expressed, although the expression of the most 3' gene HOX-4G is very low. This complex pattern, illustrated in Figure 1, is specific for the kidney. The same analysis performed in normal colon (Figure 3), lung or liver shows different patterns of expression with different HOX genes switched on or off in these organs. HOX genes of the paralogous groups 5 and 10, which seem to represent boundaries that define genes 3' and 5' displaying different expression patterns, are actively expressed in the normal kidney (see Figure 1).

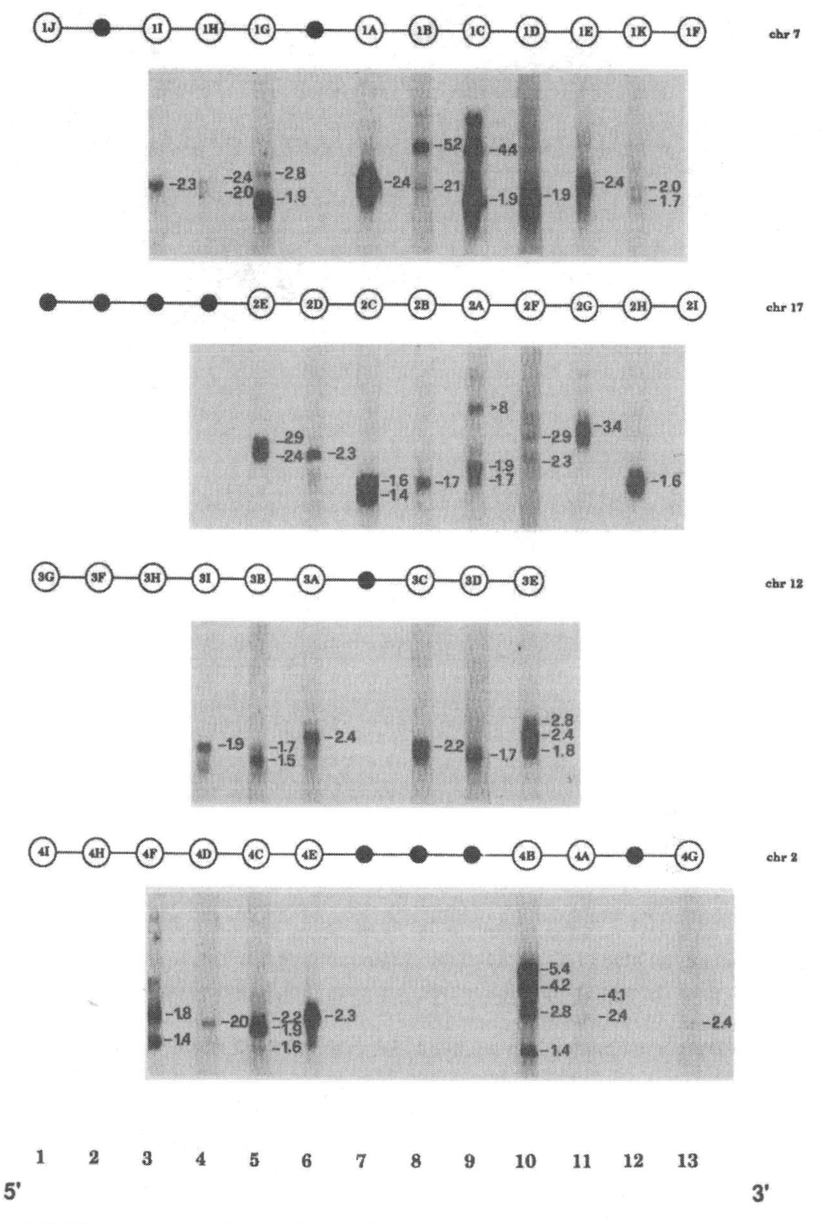

Figure 1. HOX gene expression in normal human kidney.
HOX genes are aligned horizontally according to their physical position on the chromosomes and vertically on the basis of maximal sequence homology of the homeodomain. Small stippled circles indicate homeodomains predicted in the scheme but not yet found. Northern blots of 5 μg of poly (A)+ RNA from normal kidney were hybridized to the probes indicated in the circles above each lane. Transcript sizes are given in Kb. The 13 HD groups are indicated at the bottom. (from Cillo et al. 1992)

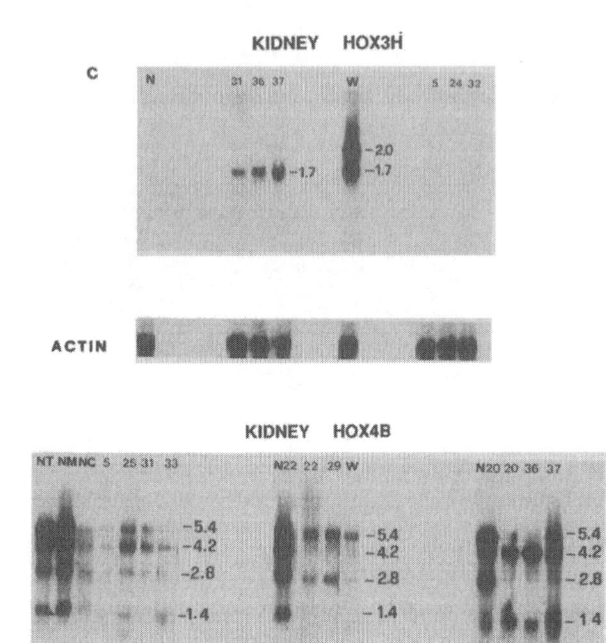

Figure 2a, 2b. Expression of HOX-2A; HOX-2E; HOX-3H and HOX-4B in normal and neoplastic kidney. Expression of the gene HOX-2A(a); HOX-2E(b); HOX-3H(c) and HOX-4B(d) in normal kidney (total = NT, medullar = NM, cortical = NC) and in kidney cancer (the numbers of the different biopsies are indicated above each lane). 5 μg of poly (A)+ RNA from normal and neoplastic kidney were electrophoresed through a 1.25% agarose gel, transferred to nitrocellulose filtres and hybridized with probes representing the 3' untranslated region of HOX-3H. Transcript size is indicated in Kb. Control hybridization to a ß-actin probe is shown. (from Cillo et al. 1992)

Major differences in the expression of HOX genes between normal kidney and renal carcinomas are detected. Three HOX-2A transcripts, of size 1.7 Kb, 1.9 Kb and greater than 8 Kb, are present in normal kidney (Figure 2a). The majority of kidney tumors tested (8/12) do not show detectable expression of this gene. However, two specimens have the same expression as normal kidney and two other biopsies show a specific expression pattern with the absence of the 1.9 molecular weight band in the only case of Wilms tumor studied and three extra high molecular weight mRNA in the other biopsy.

Two HOX-2E transcripts, of size 2.4 and 2.9 Kb are expressed in normal medullar kidney, whereas only the lower mRNA is detectable in cortical kidney (Figure 2b). The tumor samples can be classified into four categories, according to the expression of this gene: 3/12 of the samples exhibit the same pattern as normal medullar kidney; 3/12 express high steady-state levels of the 2.9 Kb transcript whereas the 2.4 Kb species is undetectable; conversely in 2/12 tumors only the lower molecular weight transcript is evident; in 4/12 tumors no HOX-2E transcripts are detected. The last category of specimens are all histologically classified as clear cell kidney carcinomas.

Expression of the HOX-3H gene is undetectable in the normal kidney and in 9/13 biopsies examined. In contrast, we have observed a 1.7 Kb HOX-3H transcript in another 3 specimens (Figure 2c). In the single Wilms tumor tested, two abundant HOX-3H transcripts of sizes 1.7 and 2.0 Kb are observed.

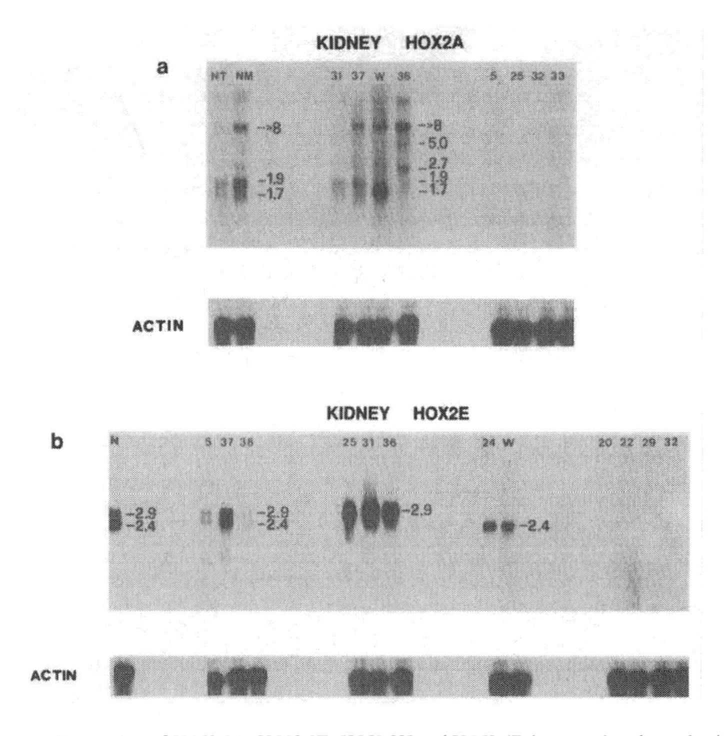

Figure 2c, 2d. Expression of HOX-2A; HOX-2E; HOX-3H and HOX-4B in normal and neoplastic kidney. See preceding page for legend.

HOX-4B is expressed in normal kidney with 4 transcripts of size 5.4, 4.2, 2.8, 1.4 Kb (Figure 2d). The molecular mechanisms regulating the expression of this gene have been well elucidated (23). Two alternative promoters underlie the transcription of two classes of HOX-4B specific mRNAs: the 5.4 and 2.8 kb transcripts are driven from a distal promoter while the 4.2 and 1.4 Kb transcripts originate from a proximal promoter. The two promoters are differentially regulated in a tissue and stage specific manner and respond differentially to retinoic acid induction (24). The expression of HOX-4B in renal carcinomas is shown in Figure 2d. On the basis of the expression of this gene it is possible to group kidney cancer into three categories: a) tumors exhibiting the same expression pattern as the normal kidney (7/13) with four major transcripts of 5.4, 4.2, 2.8 and 1.4 Kb; b) a category of tumors (3/13) with HOX-4B transcripts of size 5.4 and 2.8 Kb, shows the transcript classes presumably originating from the distal promoter, and c) 3 tumors where HOX-4B expression is presumably due to the activity of the proximal promoter, with the transcript sizes of 4.2 and 1.4Kb. It is interesting to note that normal kidney from the same patients showing the alternative transcripts presents the full spectrum of HOX-4B gene expression with the four transcript classes.

To determine whether HOX genes are differentially expressed in different parts of the kidney, we have analyzed the expression of some of the HOX genes in the cortical and medullar kidney. As shown in Figure 2d no differences in the expression patterns of the HOX-4B gene are observed; however, the levels of HOX-4B transcripts are higher in the medulla than in the cortical kidney. We have obtained the same results for the HOX-2A (Figure 2a) and HOX-3H genes (Figure 2c).

HOX GENE EXPRESSION IN PRIMARY AND METASTATIC COLORECTAL CANCER

Colorectal carcinomas rank high among the most frequent human malignancies. Such tumors may arise from benign adenomatous polyps, which later progress to better adenocarcinomas through several mutational steps (25). Some of these events have been understood through the identification of the genes Familial Adenomatous Polyposis (FAP) and Deleted in Colorectal Cancer (DCC) involved in colon tumorigenesis (26,27). The overall biological characteristics of colorectal cancers, and of neoplastic tissues in general, result from accumulated genetic alterations rather than from the order in which these events occur with respect to one another (25). Even though several important genes have been identified, other events, which remain to be elucidated, may well take place during the progression of colon cancer.

We analyzed, by Northern blotting, the expression of the 4 HOX gene clusters in normal human intestinal mucosa and in tumor samples derived from patients with primary colorectal cancer (28).

In the normal colon (Figire 3), HOX genes are highly expressed with respect to both the number of HOX genes turned on and the abundance of individual HOX transcripts.

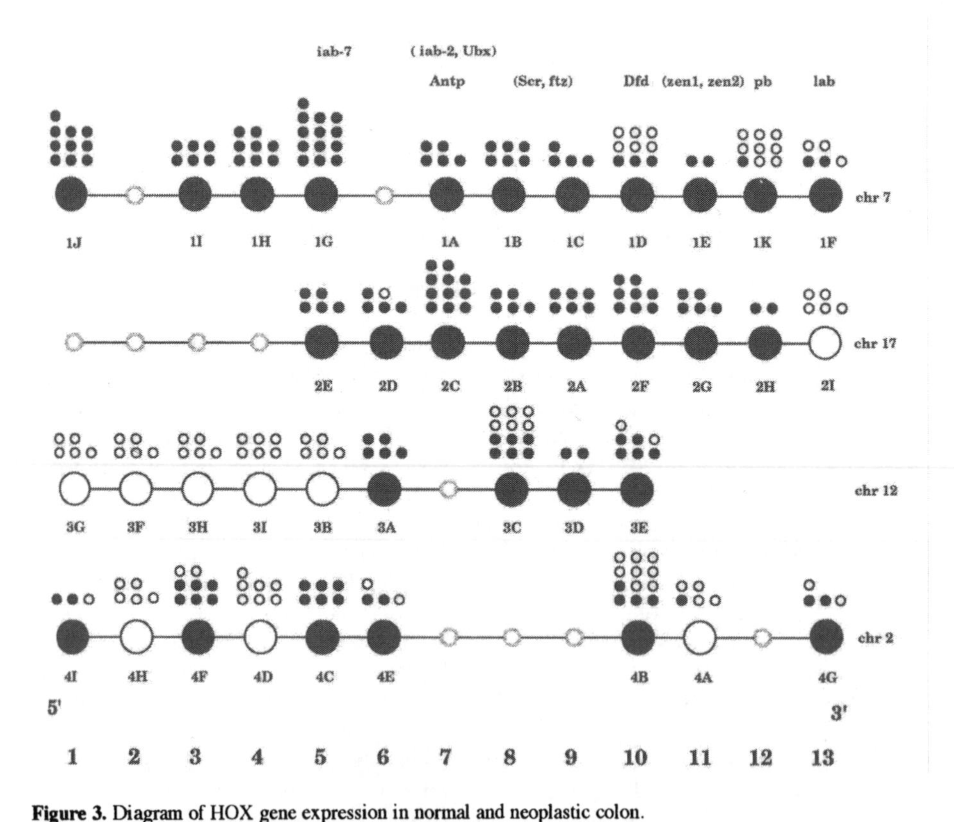

Figure 3. Diagram of HOX gene expression in normal and neoplastic colon.
Different symbols indicate different tissues. Closed or open symbols indicate active or silent HOX genes. (●○) represent normal colon; (●○) represent each individual cancer biopsy analyzed. (from De Vita et al. 1993)

Interestingly, the genes of the HOX1, HOX2 and HOX3 clusters are turned on or off in blocks, containing a variable number of contiguous genes within each locus. In contrast, contiguous genes of the HOX4 cluster are switched on or off in an alternate fashion. The expression pattern of the HOX4 cluster is peculiar to the colon since in the hematopoietic system as well as in the normal kidney HOX4 genes are coordinately expressed. The coordinate regulation of HOX genes is consistent with the idea that one or more upstream promoter elements account for the concerted expression of HOX genes in differentiating systems. Experimental evidence for a major promoter upstream of several HD containing exons of the HOX-3 locus has been reported (29).

Comparing different normal human organs such as liver, colon, kidney and lung suggests that HOX genes may display overall patterns of expression that are characteristic for each organ. Furthermore the expression of several specific HOX genes appears to vary among individual organs. Thus, in addition to their probable organ specific function, the HOX genes and their expression may become a means to molecularly fingerprinting different normal adult tissues.

Individual biopsy samples of primary tumors are markedly heterogeneous according to the expression patterns of HOX genes located at the 3' end of the HOX-1 locus. For example, HOX-1 genes from 1D through 1F, are consistently expressed in normal tissue, whereas they are silent in most colorectal cancer specimens of our series (Figure 3). Indeed, transgenic mice overexpressing the homebox gene Hox 1.4 (the murine analog of the human gene HOX-1D) exhibit abnormal gut development, resulting in an inherited abnormal megacolonic phenotype (30) reminiscent of human congenital Hirschsprung's disease. Such discontinuous turning off of the same HOX genes in different colorectal cancer specimens also concerns genes located at the 3' end of the locus HOX-3 and on the entire HOX-4 locus. Taken together, our results on primary colorectal cancer suggest that HOX genes can be regulated at transcriptional as well as at a post-transcriptional level, possibly by differential splicing.

In colorectal cancer the expression of individual HOX genes is sometimes altered as compared to that seen in normal colonic mucosa. For instance, the HOX-2C and HOX-4B

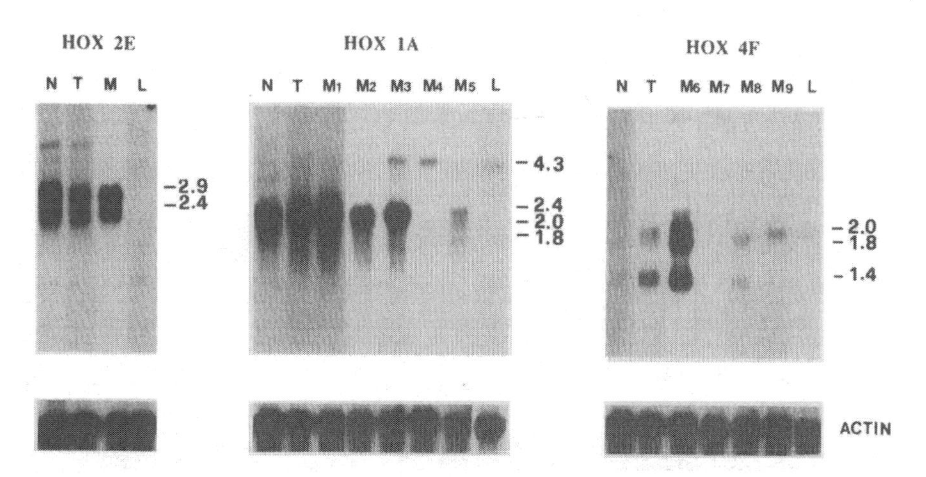

Figure 4. Expression of HOX-2E, HOX-1A, HOX-4F in normal and neoplastic colon.Expression of HOX-2E, HOX-1A and HOX-4F genes in normal colonic mucosa (N) in primary colorectal cancer (T) in normal liver (L) and in liver metastases originated from colorectal cancer (M). Transcript size is indicated in Kb Control hybridization to a ß-actin probe is shown. (from De Vita et al. 1993)

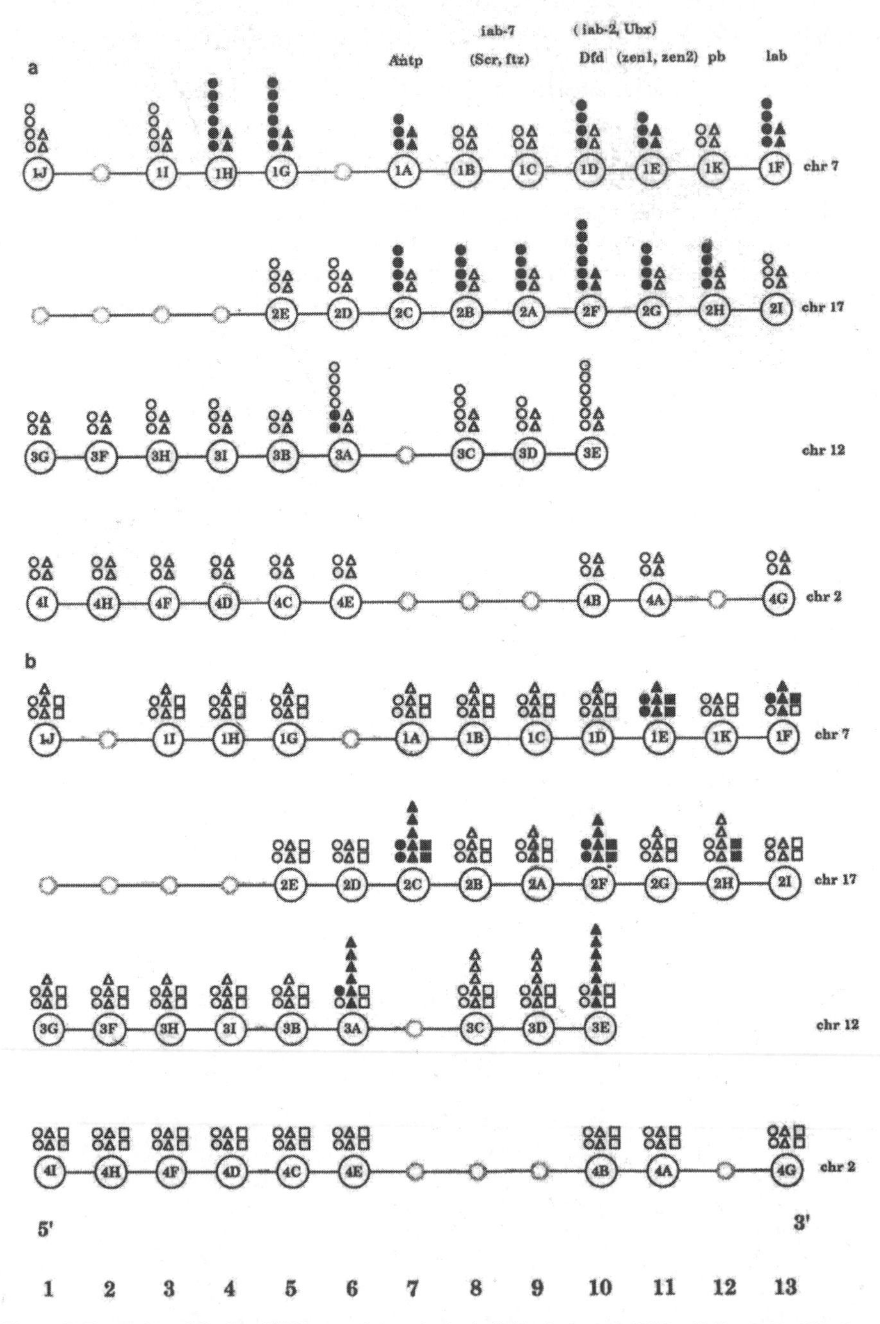

Figure 5. Expression of the four HOX gene clusters in the individual cases of ANLL, CML, pre-B-ALL, B-CLL and T-ALL.Different symbols indicate different types of leukemia. Each symbol represents each individual case analyzed within the same diagnostic category. Closed or open symbols indicate active or silent HOX genes.**Fig. 5a**: closed or open circles (●○) represent HOX genes expressed or silent, respectively, in acute nonlymphocytic leukemia (ANLL); closed or open triangles (▲△) represent HOX genes expressed or silent, respectively, in chronic myelocytic leukemia (CML) in the chronic phase.

(continued on following page)

genes, although both expressed in normal and neoplastic colon, display different classes of transcripts according to their size. For the HOX-2C gene, two transcripts with high molecular weight, already described in teratocarcinoma cells activated by retinoic acid (31), are detected in primary tumors versus normal colon. In normal colon and in 25% of the primary colon cancer biopsies, the HOX-4B gene exhibits the two transcripts driven from the distal promoter. The majority of colorectal cancer specimens do not express this gene. However, in only 1 out of the 12 cancer biopsies tested, we did observe a switching-off of the distal promoter and a switching-on of the proximal one driving two alternative transcripts.

Some HOX genes (HOX-1A, HOX-4F) (Figure 4) exhibit altered expression in metastatic lesions compared to primary colorectal tumors and to normal intestinal mucosa. Novel transcripts can indeed be expressed in some metastases compared to primary tumors or different amounts of mRNAs may be produced in metastatic versus primary colon cancer. However it has not been possible to identify any specific alterations of HOX gene expression that would systematically characterize metastatic vs. primary colorectal cancer tissues. In contrast, other HOX genes (HOX1J, HOX2E) (Figure 4) display identical patterns of expression in normal colon, primary colorectal cancer and corresponding liver metastases when these are derived from one individual patient. If this holds true with large series, it may perhaps become possible to use HOX genes as molecular probes to determine the site of a primary cancer when dealing with probable metastases of unknown primary origin.

HOX GENE EXPRESSION IN DIFFERENT TYPES OF HUMAN LEUKEMIA

The hematopoietic system is organized in a developmental hierarchy which provides an ideal model to study molecular mechanisms underlying cell proliferation and differentiation.

We have analyzed the expression of HOX genes in different types of human leukemia to investigate whether the physical organization of HOX loci reflects a regulatory hierarchy involved in the differentiation of hematopoietic cells or whether HOX gene expression might contribute to the leukemic phenotype. Patterns of HOX gene expression, by Northern blotting, were studied in several classes of myeloid and lymphoid leukemias representing cells of different lineages and stages of differentiation. Peripheral blood cells of patients with acute nonlymphocytic leukemia (ANLL) provided an enriched source of immature cells whereas cells of chronic myeloid leukemia (CML) in the chronic phase represented more differentiated populations within the myelo-monocytic pathway. We also analyzed the immature B and T cells of acute lymphatic leukemias (pre-B-ALL and T-ALL), as well as more mature B and T lymphocytes of chronic lymphocytic leukemias (B-CLL and T-CLL).

Figure 5. (continued from preceding page).
Fig. 5b: closed or open circles (●○) represent HOX genes expressed or silent, respectively, in pre-B acute lymphoblastic leukemia (pre-B-ALL); closed or open triangles (▲△) represent HOX genes expressed or silent, respectively, in chronic lymphocytic leukemia; closed or open squares (■□) represent HOX genes expressed or silent, respectively, in T-acute lymphoblastic leukemia (T-ALL).
HOX genes are aligned horizontally according to their physical position on the chromosomes and vertically on the basis of maximal sequence homology of the homeodomains. Small stippled circles indicate homeodomains predicted in the scheme but not yet found. Homeodomains from *Drosophila* BX-C and ANT-C genes are indicated above the scheme. Each fly homeodomain has been placed on top of the group of human homeodomains to which it is most closely related in sequence. Correspondence of homeodomains in brackets is unclear. (from Celetti et al. 1993).

Our results, illustrated in Fig. 5, demonstrate that different types of human leukemia are characterized by specific patterns of HOX gene expression (32). Virtually identical patterns of expression of the four HOX gene clusters were observed in different patients with the same type of leukemia suggesting that it may be possible to define distinct patterns of HOX gene expression specific for each form of leukemia. Variability was observed only in the expression of one out of 38 HOX genes in ANLL and T-ALL and of two genes in B-ALL. Strikingly, this minor variability was restricted to the same genes (HOX3A and HOX1F) in different pathologies.

The second major conclusion from these experiments is that HOX genes are coordinately regulated in blocks in myeloid cells whereas in both the B and T lymphoid cell lineages they are activated in an independent fashion. There was very low, if any, expression of HOX3 and HOX4 genes whereas genes within the HOX1 and HOX2 clusters were present at much higher levels Six contiguous genes of the HOX2 locus, 2C through 2H, active in ANLL blasts, were not expressed in CML cells in the chronic phase. Furthermore two adjacent genes 5' of HOX1, expressed at high levels in ANLL, were expressed at much lower levels in CML cells. Only one case of CML in acute phase of different phenotype was analyzed and it is clear that more cases need to be analyzed before definitive conclusions can be reached. Nevertheless it is striking the observation that the patterns of HOX gene expression in patients with CML in myeloid or lymphoid blast crisis were considerably different from one another and from patients with CML in chronic phase. Furthermore, the expression pattern of CML in myeloid blast crisis was almost identical to the pattern observed in ANLL while HOX gene expression in lymphoid blast crisis was very similar to that observed in B-ALL.

The expression of sets of HOX2 genes is associated with early stages of differentiation. As cells maturate along the granulocyte-macrophage pathway, HOX2 expression is then down-regulated. These observations are consistent with previous results showing that eight out of nine genes of the HOX2 locus are active in erythroleukemic cell lines but not in myelomonocytic cells (18). Perkins *et al.* have suggested that expression of HOX2.4 gene impedes the terminal differentiation of myeloid cells (14). Our results are consistent with this possibility and furthermore suggest that down-regulation of sets of HOX2 genes might be required for cell maturation.

It is interesting to contrast the pattern of HOX gene expression in lymphoid and myeloid cells. Whereas HOX genes are expressed as a block in myeloid cells they appear to function as isolated genes in lymphoid cells. Only two or three non-contiguous genes of the HOX2 locus (2C, 2F and 2H) were expressed in lymphoid blasts of the B and T cell lineages and a few scattered genes of the HOX1 (1A and 1F) and HOX3 clusters (3A and 3E) were active in mature T and B lymphocytes, respectively. These observations suggest that HOX genes are regulated differently in myeloid and lymphoid cells. Finally we analyzed HOX gene expression in B cells from tonsils and in $CD3^+$ peripheral blood cells representing the normal counterparts of B and T cell populations. Virtually identical results were obtained in normal and leukemic populations. Normal B lymphocytes, as B-CLL cells, exhibited high levels of expression of the HOX3E gene, and, at lower intensity, of HOX3A. The only difference was the expression of HOX1A and HOX2H in B-cells which was not detected in B-CLL.

It is interesting that the HOX4 locus was never expressed in either normal or leukemic cell populations, suggesting that this cluster is not involved in hematopoiesis. On the contrary, the most significant variations in HOX gene expression between the different types of leukemia occur within the HOX2 cluster.

CONCLUSIONS

HOX gene clusters display characteristic patterns of expression in normal organs indicating that the coordinate regulation of HOX genes may play an important role in normal organogenesis. The alteration of HOX gene expression observed in kidney and colon cancers seems to stress that transcription factors play an important role in cancer evolution. Further work is needed to assess the correlation between alterated HOX gene expression and cancer. The identification of target genes that are activated or repressed by the HOX genes will facilitate our understanding of the molecular mechanisms involved in neoplastic transformation.

The study on HOX gene expression in leukemic cell populations suggest that hematopoietic cells express a repertoire of HOX genes that is characteristic of a particular cell lineage at a specific stage of differentiation. These characteristic patterns of HOX gene expression may reflect the potentially important role that these genes play in cell lineage determination in both normal and leukemic hematopoiesis. It is therefore not surprising that no major change in the pattern of HOX gene expression was observed in leukemic cells compared to their normal counterparts. The characteristic overall patterns of HOX gene expression might provide a means of "fingerprinting" cells with respect to their type and stage of differentiation. It is also possible that alterations in the expression of individual HOX genes may directly contribute to leukemic transformation. Therefore analysis of HOX gene expression in populations of normal progenitors within the hematopoietic stem cell hierarchy might provide insights into the role that individual genes within this large family play in hematopoiesis.

ACKNOWLEDGMENTS

This work was supported by the Italian Association for Cancer Research (AIRC), the Consiglio Nazionale delle Ricerche (CNR) progetti finalizzati Biotecnologia e Biostrumentazione and Ingegneria Genetica.

REFERENCES

1. Weinberg, R.A., Oncogenes, antioncogenes, and the molecular bases of multistep carcinogenesis. *Cancer Res.*, 49:3713-3721 (1989).
2. Varmus, H.E., Oncogenes and transcriptional control. *Science*, 218:1337-1339 (1987).
3. Distel, R.J., Ro, H.S., Rosen, B.S., Groves, D.L. and Spiegelman, B.M., Nucleoprotein complexes that regulate gene expression in adipocyte differentiation: direct participation of c-fos. *Cell*, 49:835-844 (1987).
4. Bohmann, D., Bos, T.J., Admon, A., Nishimura, T., Vogt, P.K., Tjian, R., Human proto-oncogene c-jun encodes a DNA binding protein with structural and functional properties of transcription factor AP-1. *Science*, 238:1386-1392 (1987).
5. Green, S. and Chambon, P., A superfamily of potentially oncogenic hormone receptors. *Nature*, 324: 615-617 (1986).
6. Chellappan, S.P., Hiebert, S., Mudryj, M., Horowitz, J.R., Nevins, J.R., The E2F transcription factor is a cellular target for the RB protein. *Cell*, 65:1053-1061 (1991).
7. Gehering, W.J., and Hiromi Y., Homeotic genes and the homeobox. *Ann. Rev. Genet.*, 20:147-173 (1986).
8. Han K., Levine M.S., and Manley J.L., Synergistic acitivation and repression of transcription by *Drosophila* homeobox proteins. *Cell*, 56:573-583 (1989).
9. Levine M., Rubin G.M., and Tjian R. Human DNA sequences homologous to a protein coding region conserved between homeotic genes of *Drosophila*. *Cell*, 38:667-673 (1984).

10. Akam, M., HOX and HOM: Homologous gene clusters in insects and vertebrate. *Cell*, 57:347-349 (1989).
11. Acampora, D., D'esposito M., Faiella A., Pannese M., Migliaccio E., Morelli F. et al. The human HOX gene family. *Nucl. Acids Res.*, 17:10385-10402 (1989).
12. Graham A., Papalopulu N., and Krumlauf R. The murine and *Drosophila* homeobox gene complexes have common features of organization and expression. *Cell* , 57:367-378 (1989).
13. Simeone A., Mavilio F., Acampora D., Giampaolo A., Faiella A., Zappavigna V., et al. Two human homeobox genes C1 and C8: structure, analysis and expression in embryonic develompent. *Proc. Natl. Acad. Sci. USA*, 84:4914-4918 (1987).
14. Perkins, A., Kongsuwan, K., Visvader, J., Adams, J.M. and Cory, S., Homeobox gene expression plus autocrine growth factor production elicits myeloid leukemia. *Proc. Natl. Acad. Sci. USA*, 87, 8398-8402, (1990).
15. Kamps M.P., Murre C., Sun X-H., and Baltimore D. A new homeobox gene contributes the DNA binding domain of the t(1; 19) translocation protein in pre-B ALL. *Cell*, 60:547, 555, (1990).
16. Hatano M., Roberts C.W.M., Minden M., Crist W.M., Korsmeyer S.J.: Deregulation of a homeobox gene, HOX11, by the t(1;14) in T cell leukemia. *Science* 253:79-82, (1991).
17. Cillo C., Barba P., Bucciarelli G., Magli M.C., and Boncinelli E.. Hox gene expression in normal and neoplastic kidney. *Int. J. of Cancer* 51:892-897 (1992).
18. Magli, M.C., Barba, P., Celetti, A., De Vita, G., Cillo, C. and Boncinelli, E. Coordinate regulation of HOX genes in human hematopoietic cells. *Proc. Natl. Acad. Sci. USA*, 88:6348-6352 (1991).
19. Saxen, L., Organogenesis of the Kidney. *Cambrige University Press* (1987).
20. Simeone, A., Acampora, D., Nigro, V., Faiella, A., D'Esposito, M., Stornaiuolo, A., Mavilio, F. and Boncinelli E. Differential regulation by retinoic acid of the homeobox genes of the four HOX loci in human embryonal carcinoma cells. *Mech. Dev.*, 33:215-228 (1991).
21. Stornaiuolo, A., Acampora D., Pannese M., D'Esposito M., Morelli F., Migliaccio E., et al. Human HOX genes are differentially activated by retinoic acid in embryonal carcinoma cells according to their position within the four loci. *Cell Diff. Dev.*, 31:119-127 (1990).
22. D'Esposito, Morelli F., Acampora D., Migliaccio E., Simeone A., and Boncinelli E. EVX2, a Human homeobox gene homologous to the even-skipped segmentation gene, is located at the 5' end of HOX4 locus on chromosome 2. *Genomics* , 10:43-50 (1991).
23. Cianetti, L., Di Cristoforo, A., Zappavigna, V., Bottero, L., Boccoli, G., Testa, U., Russo, G., Boncinelli, E. and Peschle, C., Molecular mechanisms underlying the expression of the human HOX-5.1 gene. *Nucleic Acids Res.*, 18:4361-4368 (1990).
24. Mavilio, F., Simeone, A., Boncinelli, E., and Andrews, P.W. Acitivation of four homeobox gene clusters in human embryonal carcinoma cells induced to differentiate by retinoic acid. *Differentiation*, 37:73-79 (1988).
25. Fearon, E.R. and Vogelstein B. A genetic model for colorectal tumorigenesis. *Cell* , 61:759-767 (1990).
26. Leppert, M., Dobbs M., Scambler P., O'Connell P., Nakamura Y., Stauffer D., et al. The gene for familial polyposis coli maps to the long arm of chromosome 5. *Science* , 238:1411-1413 (1987).
27. Fearon, E.R., Cho K.R., Nigro J.M., Kern S.E. Somons J.W., Ruppert J.M., et al. Identification of a chromosome 18q gene that is altered in colorectal cancers. *Science*, 247:49-56 (1990).
28. De Vita, G., Barba P., Odartchenko N., Givel J.C., Freschi G., Bucciarelli G., Magli M.C., Boncinelli E. and Cillo C. Expression of homeobox-containing genes in primary and metastatic colorectal cancer. *Eur. J. of Cancer* 29A:887-893 (1993).
29; Simeone A., Acampora D., D'Esposito M., Faiella A., Pannese M. and Boncinelli E. At least three human homeoboxes on chromosome 12 belong to the same transcription unit. *Nucleic Acids Res*, 16:5379-5387 (1988).
30. Wolgemuth D.J., Behringer R.R., Mostoller M.P., Brister R.L. and Palmiter R.D. Transgenic mice overexpressing the mouse homeobox-containing gene HOX-1.4 exhibit abnormal gut development. *Nature* , 337:464-467 (1989).

31. Simeone A. Acampora D., Nigro V., Faiella A., D'Esposito M., Stornaiuolo A., et al. Differential regulation by retinoic acid of the homeobox genes of the four HOX loci in human embryonal carcinoma cells. *Mech. Dev*, 33:215-228 (1991).
32. Celetti, A., Barba P., Cillo C., Rotoli B., Boncinelli E. and Magli M.C. Characteristic patterns of HOX gene expression in different types of human leukemia. *Int. J. of Cancer*. In press.

THYROID SPECIFIC EXPRESSION OF THE KI-RAS ONCOGENE IN TRANSGENIC MICE

Giovanni Santelli[1], Vittorio de Franciscis[2], Gennaro Chiappetta[1], Amelia D'Alessio[2], Daniela Califano[1], Alba Mineo[1], Carmen Monaco[1], and Giancarlo Vecchio[2]*

[1]Istituto per lo Studio e la Cura dei Tumori
Fondazione "G. Pascale"
Via M. Semmola
80131 Napoli, Italy
[2]Centro di Endocrinologia ed Oncologia Sperimentale del C.N.R.
c/o Dipartimento di Biologia e Patologia Cellulare e Molecolare
Università di Napoli
Via S. Pansini, 5
80131 Napoli, Italy

INTRODUCTION

Few oncogenes have been found consistently activated in the DNA from human naturally occurring carcinomas of the thyroid gland: ras (Ki-, Ha-, N-ras), trk and ret/PTC (1-4). Among them the trk and ret/PTC oncogenes are found activated exclusively in the papillary histotype (4). In contrast ras oncogenes are frequently found activated in all thyroid carcinoma histotypes ranging from the benign adenoma to the clear malignant adenocarcinoma, thus suggesting that ras activation is an early step in thyroid tumorigenesis (5,6). Different experimental models of thyroid neoplasias have been obtained with ras oncogenes, either by transforming thyroid epithelial cells in vitro or by directly injecting the Kirsten murine sarcoma virus (KiMSV) into the rat thyroid gland (7-10). It has been demonstrated that the Ha-ras is not able to fully transform human thyroid epithelial cells (10), and we have previously demonstrated that ras oncogenes are able to transform the thyroid epithelial cell line PC c13, only in the presence of another cooperating oncogene (8). To further characterize the role of ras oncogenes in the development of thyroid carcinomas we have generated transgenic mice in which the expression of the Ki-ras oncogene is specifically addressed to the thyroid gland. For this purpose transgenic animals carrying constructs composed by segments of the rat thyroglobulin (TG) (11) gene promoter and by a human mutated Ki-ras oncogene cDNA (12) have been obtained. One thyroid adenoma with multiple

*To whom requests for reprints should be addressed.

Advances in Nutrition and Cancer, Edited by
V. Zappia *et al.*, Plenum Press, New York, 1993

foci has appeared after a long latency period in one transgenic mouse. This indicates that the ras oncogene is not able by itself to induce a full malignant transformation of the mouse thyroid.

GENERATION OF TRANSGENIC MICE

The rat TG gene 5' flanking sequence (spanning from -1950bp to +39bp) was fused to the entire coding sequence of the human Ki-ras cDNA, carrying the activating aminoacid conversion in the codon 12 (12). The construct was microinjected into one cell mouse embryos, then reimplanted in a foster mother, and the progeny analyzed for the integration of the transgene (13). We have obtained 17 founder mice with integrated construct (TG2000ras). All founders gave rise to lines of transgenic animals. Animals belonging to the Fl generation were sacrified at different ages between 3 and 12 months. The thyroid gland and other control tissues (brain, liver, lung, and kidney) were isolated, frozen sections were fixed in paraformaldehyde, then stained with ematoxylin and eosin. The large majority of thyroids analyzed from transgenic animals, as well as the totality of the control tissues from the same animals, have no apparent anomalies or pathological finding. We have shown clear anomalies affecting the thyroid gland of two transgenic mice: a) the founder TG2000-40 presented a marked vasculitis, and b) TG2000-40.4, an Fl mouse from the same family, showed a thyroid histologically abnormal, with various lesions that can be classified as adenomas and nodules. Table 1 summarizes the histology of the examined mice.

Table 1. Histology of TG2000-ras transgenic mice.

MOUSE	SEX	AGE (months)	HISTOLOGY
$F_0$4	M	21	NORMAL
$F_1$2	F	13	NORMAL
$F_2$1	F	12	NORMAL
$F_2$2	F	12	NORMAL
$F_0$21	M	20	NORMAL
$F_1$5	M	7	NORMAL
$F_1$15	F	6	NORMAL
$F_0$35	M	15	NORMAL
$F_1$3	F	22	NORMAL
$F_1$4	F	22	NORMAL
$F_0$40	F	16	VASCULITIS
$F_1$1	M	12	ADENOMA
$F_1$7	F	15	NORMAL
$F_1$9	F	15	NORMAL
$F_1$11	F	9	NORMAL
$F_1$12	F	15	NORMAL
$F_0$61	F	15	N. D.
$F_1$9	M	18	NORMAL
$F_1$11	M	14	NORMAL
$F_1$12	M	17	NORMAL

TISSUE EXPRESSION OF THE CONSTRUCT

By using in situ mRNA hybridization we have analyzed tissue slices derived from the thyroid glands of 8 transgenic and 8 control non-transgenic mice. The ribo-probes (anti-sense or sense) were synthesized in vitro in presence of ^{35}S labeled precursors by using as template the same human Ki-ras sequences used to construct the transgene. These experiments are still in progress, however it seems that ras is expressed in two out of five families analyzed. On the other hand it is well known that ras malignant transformation of a rat thyroid cell line in vitro is followed by the block of thyroglobulin gene expression, as well as the other thyroid specific functions (7). Thus it is not possible to exclude that the low tumor incidence that we found in these animals could be also partially attributed to the low levels of expression of the transgene in the thyroid gland.

CONCLUSIONS

The use of transgenic mice is a powerful tool to investigate in vivo oncogene action during transformation. Tissue-specific expression of activated ras oncogenes in transgenic mice results in tumor appearence with variable incidence (15-18), and, in some cases, the ras transgene was also able to cooperate with other oncogene products to transform the target mouse tissue (19). In this regard the thyroid gland provides a unique model to investigate multistep carcinogenesis and the cooperation among oncogenes. In fact, spontaneous thyroid tumors in mice occur very rarely (ref. 14 and our own results of autopsies performed on more than 50 control animals). In this context our data, i.e. the appearence of benign tumors with a long latency, are in good agreement with literature data on thyroid transformation. In fact the expression of the activated Ki-ras oncogene seems not to be a sufficient event to make the thyroid epithelial cells malignant in vivo. Additional genetic events (possibly trk or ret/PTC activation) seem in fact to be required for full transformation of the thyroid gland. This mouse model will enable us to study the thyroid transformation process under the influence of enviromental or physiological stimuli and the role played by the action of other oncogenes during this process.

ACKNOWLEDGEMENTS

Supported in part by the Associazione Italiana Ricerca sul Cancro, Milan, Italy, by the CNR Target Project "Biotechnology and Bioinstrumentation", and by the CNR Target Project "Applicazioni Cliniche della Ricerca Oncologica". A.D'A. is a recipient of a fellowship from CNR Target Project "Biotechnology and Bioinstrumentation".

REFERENCES

1. Suarez, H.G., du Villard, J.A., Séverino, M., Caillou, B., Schlumberger, M., Tubiana, M., Parmentier, C., and Monier, R. Presence of mutations in all three ras genes in human thyroid tumors. Oncogene 5:565-570, 1990.
2. Bongarzone, I., Pierotti, M.A., Monzini, N., Mondellini, P., Manenti, G., Donghi, R., Pilotti, S., Grieco, M., Santoro, M., Fusco, A., Vecchio, G., and Della Porta, G. High frequency of activation of tyrosine kinase oncogenes in human papillary thyroid carcinoma. Oncogene 4:1457-1462, 1989.
3. Fusco, A., Grieco, M., Santoro, M., Berlingieri, M.T., Pilotti, S., Pierotti, M.A., Della Porta, G., and Vecchio, G. A new oncogene in human thyroid papillary carcinomas and their lymph-nodal metastases. Nature, 328:170-172, 1987.

4. Jhiang, S.M., Caruso, D.R., Gilmore, E, Ishizaka, Y.,Tahira, T., Nagao, M., Chiu, I., and Mazzaferri, E.L. TPC Detection of the PTC/ret oncogene in human thyroid cancer. Oncogene, 7:1331-1337, 1992.
5. Lemoine, N.R., Mayall, E.S., Wyllie, F.S, Williams, E.D.Goyns,M., Stringer, B., and Wynford-Thomas, D. High frequency of ras oncogene activation in all stages of human thyroid tumorigenesis. Oncogene, 4:159-164, 1989.
6. Wright, P.A., Lemoine, N.R., Mayall, E.S., Wyllie, F.S., Hughes, D., Williams, E.D., and Wynford-Thomas, D. Papillary and follicular thyroid carcinomas show a different pattern of ras oncogene mutations. Br.J.Cancer, 60:576-577, 1989.
7. Vecchio, G., Di Fiore, P.P., Fusco, A., Colletta, G., Weissman, B.E., and Aaronson S.A. In vitro tranformation of epithelial cell by acute retroviruses. In: F.Blasi (ed.), Human Genes and Diseases, pp. 415-470. John Wiley & Sons, Ltd., 1986.
8. Fusco, A., Berlingieri, M.T., Di Fiore, P.P., portella, G., Grieco, M., and Vecchio, G. One- and two-step transformation of rat thyroid epithelial cells by retroviral oncogenes. Mol. Cell. Biol, 7:3365-3370, 1987.
9. Portella, G, Ferulano, G., Santoro, M., Grieco, M., Fusco, A., and Vecchio, G. The Kirsten murine sarcoma virus induces rat thyroid carcinomas in vivo. Oncogene, 4:181188, 1989.
10. Lemoine, N.R., Staddon, 5., Bond, J., Wyllie, F.S., Shaw, J.J., and Wynford-Thomas, D. Partial transformation of human thyroid epithelial cells by mutant Ha-ras oncogene. Oncogene, 5:1833-1837, 1990.
11. Musti, A.M., Ursini, V.M., Aw edimento, E.V., Zimarino, V., and Di Lauro, R. A cell type specific factor recognizes the rat thyroglobulin promoter. Nucl. Acids Res., 15:81498166, 1987.
12. McCoy, M., Bargmann, C.L., and Weinberg, R.A. Human colon carcinoma Ki-ras 2 oncogene and its corresponding proto-oncogene. Mol. Cell Biol., 4:1577-1582, 1984.
13. Hogan, B., Costantini, F., and Lacy, E. Manipulating the mouse embryo- A laboratory manual. Cold Spring Harbor Laboratory, Cold Spring Harbor, 1986.
14. Biancifiori, C. Pathology of tumors in laboratory animals. Vol.II, Tumors of the mouse. IARC Scientific Publication n 23, Ed. in chief V.S. Turousov, Lyon, 1979, 451-463
15. Quaife, C.J., Pinkert, C.A., Ornitz, D.M., Palmiter,R.D., and Brinster, R.L. Pancreatic neoplasia induced by ras expression in acinar cells of transgenic mice. Cell, 48:1023-1034, 1987.
16. Andres, A.C., Schonenberger, C.A., Groner, B., Henninghausen, L., LeMeur, M., and Gerlinger P. Ha-ras oncogene expression directed by a milk proteingene promoter: tissue specificity, hormonal regulation, and tumor induction in transgenic mice. Proc.Natl.Acad.Sci. USA, 84:1299-1303, 1987.
17. Bailleul, B., Surani, M.A., White, S., Barton, S.C., Brown, K., Blessing, M., Jorcano, J., and Balmain, A. Skin hyperkeratosis and papilloma formation in transgenic mice expressing a ras oncogene from a suprabasal keratin promoter. Cell, 62:697-708, 1990.
18. Sandgren, E.P., Quaife, C.J., Pinkert, C.A., Palmiter, R.D., and Brinster, R.L. Oncogene-induced liver neoplasia in transgenic mice. Oncogene, 4:715-724, 1989.
19. Sinn, E., Muller, W., Pattengale, P., Tepler, I., Wallace, R., and Leder, P. Coexpression of MMTV/v-Ha-ras and MMTV/c-myc genes in transgenic mice: synergistic action of oncogenes in vivo. Cell, 49:465-75, 1987

EPIDEMIOLOGICAL STUDIES: RISK FACTORS AND DIET

NUTRITION AND CANCER: GENERAL CONSIDERATIONS

Flaminio Fidanza

Istituto di Scienza dell'Alimentazione
Università degli Studi
06100 Perugia, Italy

With the International Cooperative Study on the Epidemiology of Coronary Heart Disease (more commonly known as the Seven Countries Study) we examined sixteen "chunk" samples of men aged 40 to 59, resident for more than five years in each area.

For this longitudinal study, the standardized examination procedure included questionnaires on family status and medical history, anthropometry, physical examination, ECG, blood samples, qualitative urinalysis and dietary appraisal (Keys et al. 1967).

In 1965 we examined in detail the food intake of all 1536 men aged 45-64 years of age of the two rural Italian cohorts from Crevalcore in the north near Bologna and Montegiorgio in the center near Ancona. The dietary history method was used (Alberti Fidanza et al., 1988) and the two cohorts were followed for the next 20 years for total and specific mortality (Farchi et al., 1989).

All subjects were classified into four different groups depending on the nutrient densities of their diet, using a k-means cluster analysis technique. In Table 1 the mean values of food consumption in each cluster are shown.

Cluster 1 presents the highest intake of alcohol and the lowest intake of most other foods. Cluster 2 has the highest consumption of seed oils and a rather low intake of olive oil, sausages, fish, vegetables and eggs. Cluster 3 presents highest intake of meat, sausages, fish, vegetables, cheese, olive oil, other fats, and the lowest of starchy foods. Cluster 4 is characterized by the highest intake of starchy foods and vegetables and an intermediate of oils. Alcohol intake is not high and rather similar for cluster 2, 3 and 4. The diet of subjects from cluster 4 can be considered a moderate mediterranean diet, typical of the working class in Italy in the fifties. In Table 2 the age-adjusted death rates for specific cancers at 20 year follow-up are shown.

Cluster 2, which presents the highest rate for stomach cancer, has the highest intake of polyunsaturated fatty acids. This rate remains high after exclusion of the most prevalent cases in the first five years. Lung cancers do not seem related to diet and for other specific cancers the number was too small to be reported. For overall cancer mortality, cluster 1 presents the highest death rate and the trend with other groups is negatively correlated with vitamin C intake and also niacin and riboflavin although the differences among groups are small.

Advances in Nutrition and Cancer, Edited by
V. Zappia *et al.*, Plenum Press, New York, 1993

Table 1. Daily food consumption in Crevalcore and Montegiorgio (in grams).*

Food	Cluster 1	Cluster 2	Cluster 3	Cluster 4
Starchy food (bread, rice, potatoes, etc)	455±7.3	434±11.8	401±6.3	590±8.0
Vegetables	51±2.3	42±2.6	55±2.4	55±2.4
Pulses	6±0.5	5±0.7	6±1.0	6±0.6
Fruit	155±8.0	212±13.8	198±8.2	196±8.4
Fish	21±0.9	18±1.3	22±1.0	21±1.0
Seed oil	6±0.6	45±1.4	2±0.3	5±0.4
Olive oil	30±1.0	5±0.9	43±1.0	26±0.7
Other fats	22±0.7	21±1.2	26±0.9	17±0.5
Meat	77±2.8	101±4.3	105±3.5	93±3.0
Eggs	18±1.0	15±1.3	18±1.1	16±0.9
Sausages	25±1.3	17±1.6	27±1.4	21±1.1
Cheese	10±0.7	13±1.1	19±1.2	12±0.8
Cakes, biscuit etc	17±1.1	34±2.2	31±1.9	29±1.5
Alcohol	150±2.8	67±2.6	55±1.5	62±1.5

* Values taken from Farchi et al. (1989).

The alcohol intake in these population groups is rather high and so we examined in greater detail the role of this nutrient to mortality (Farchi et al., 1992). The consumption ranges from 0 to 330 g/day, with a mean value of 84±54 g/day, mostly as wine. The appropriate statistical analysis shows a J-shaped relationship between alcohol consumption, as percentage of total daily energy intake, and overall cancer mortality.

Table 2. Age-adjusted death rates for all cancers, stomach and lung cancers at 20 year follow-up in Crevalcore and Montegiorgio.*

	Cluster 1	Cluster 2	Cluster 3	Cluster 4
No.	(439)	(185)	(423)	(489)
Cause of death				
All cancers	21.7±2.3	13.6±3.0	16.0±1.9	14.8±1.7
Stomach cancer	4.1±1.2	6.0±2.1	5.0±1.2	2.6±0.9
Lung cancer	2.4±0.9	2.8±1.4	3.5±1.0	2.2±0.7

* Values taken from Farchi et al. (1989)

Slightly differing from cardiovascular disease mortality, the mortality rates in the first four quintiles of alcohol consumption do not differ significantly, but the difference is highly significant for the quintile of heavy drinkers (mean: 164.7 g/d).

From these results it seems that a Mediterranean diet can have a preventive effect also for some cancers. But in the Mediterranean countries the healthful diet of some years ago is greatly changing, approaching the North-European and American pattern.

We have already examined this aspect in detail (Fidanza, 1991). According to our study, an "Italian Mediterranean Diet Reference" can be considered one of the subjects from Nicotera, a poor rural area of Calabria in Southern Italy, included in 1960 in the Seven Countries Study, but not followed longitudinally because of shortage of money and similarity with the two rural areas of Greece. In this diet cereals were prevailing, as well as vegetables, legumes and fish. Olive oil, produced locally, was the only fat used. Meat, eggs, cheese and milk were consumed rather seldom. Wine was consumed moderately by men.

It will be difficult to convince people of our society accustomed to their particular modern life style to return to old nutrition patterns.

As stated previously (Alberti-Fidanza, 1990), we may wonder whether today it is right to categorically recommend the classical Mediterranean meal patterns. These were well suited to particular life styles, and for this reason we speak of "Mediterranean diet" not only of Mediterranean food habits. The nutritional suggestions spread dogmatically or stressing certain foods risks either being overlooked or else causing misunderstandings and strong cognitive disagreements. Consequently foods, such as bread, pasta, legumes, olive oil, could be included in the diet in an incorrect way.

Through cooperation among food industries, biotechnology and nutrition research, new solutions will have to be found as well as new nutrition education approaches which suit the needs of the individual citizen and of society.

REFERENCES

Alberti Fidanza, A., Seccareccia, F., Torsello, S., and Fidanza, F., 1988, Diet of Two Rural Population Groups of Middle-Aged Men in Italy, Int. J. Vit. Nutr. Res 58:442.

Alberti Fidanza, A., 1990, Mediterranean Meal Patterns, Bibliotheca Nutritio et Dieta 45:59.

Farchi, G., Mariotti, S., Menotti, A., Seccareccia, F., Torsello, S., and Fidanza, F., 1989, Diet and 20-y mortality in two rural population groups of midle-aged men in Italy, Amer. J. Clin. Nutr. 50:139.

Farchi, G., Fidanza, F., Mariotti, S., and Menotti, A., 1992, Alcohol and Mortality in the Italian Rural Cohorts of the Seven Countries Study, Intern. J. Epidemiol. 21:74.

Fidanza, F., 1991, The Mediterranean Italian diet: keys to contemporary thinking, Proc. Nutr. Soc. 50:519.

Keys, A., Aravanis, C., Blackburn, H.W. et al., 1967, Epidemiological studies related to coronary heart disease: characteristics of men aged 40-59 in seven countries, Acta Med. Scand: Suppl. 460.

DIET AND PRECANCEROUS LESIONS

Michael J. Hill

ECP (UK)
41 London Street
Andover
Hants, SP10 2NU
UK

INTRODUCTION

No major disease has ever been controlled by treatment; all of the successes in disease control have been achieved through prevention. Thus polio and smallpox still cannot be treated successfully and any person who suffers such virus infection will suffer to an extent similar to that experienced by patients 100 years ago. Nevertheless, both diseases have been successfully controlled by vaccination and later, in the case of smallpox, by eradication. Cholera, typhoid, botulism, dysentery etc are severe and debilitating bacterial diseases that are nevertheless controlled by public health measures. Similarly cancer will not be controlled by treatment (although mortality will, of course, be reduced) but only by prevention.

Many cancers have been associated with diet; in the analysis of causes of cancer mortality in the United States (Wynder and Gori, 1977; Doll and Peto, 1981) dietary factors have been estimated to be responsible for 30-40% of all cancer -- a similar level to that of tobacco. However, whereas the links between cancer and tobacco usage tend to be clear and the likely effects of tobacco avoidance readily estimated, the relationship between diet and cancer risk is much less clear.

In the recent ECP symposium on public education on diet and cancer (Benito et al., 1992) it was concluded that the only recommendations that could be made were to eat more fruit and vegetables and to avoid obesity. Recommendations to eat less fat or more fibre had to be suspended until the situation could be clarified further. The lack of clarity is almost certainly due to the complexity of the carcinogenesis process. Cancer is the result of a multistage process in which the causes of the individual stages may differ profoundly. This is discussed in detail elsewhere in this symposium in, for example, the chapters on cancer of the breast, the stomach and colorectum. In breast cancer there is a strong association with dietary animal fat in studies of populations (Carroll, 1985) but it is only weakly, if at all, in case-control studies (De Waard, 1992). The latter has suggested that this is because fat is only implicated in the early stages of the disease. The implication of this is that control of fat intake would only decrease breast cancer risk if undertaken in early life and not if the intervention were in the high-risk postmenopausal period of life.

Advances in Nutrition and Cancer, Edited by
V. Zappia *et al.*, Plenum Press, New York, 1993

Similarly, H. pylori infection and low intake of ascorbic acid are weakly associated with the overall process of carcinogenesis of the stomach but strongly with the precancerous stage of intestinal metaplasia and dysplasia (Reed, 1992).

In this presentation I will first discuss the relation between precursor and associated lesions, followed by some examples of known precancerous lesions at different sites, the relation between diet and these lesions and the implications of the results for cancer prevention.

PRECURSOR AND ASSOCIATED LESIONS AND DISEASES

Epidemiological studies have revealed many correlations between the risk of cancer at a particular site and the prevalence of other disease. In some cases these are simply coincidental correlations. In other cases the diseases may be correlated because they both have a common component in their etiology; such diseases or lesions are termed "associated" with neither being a precursor of the other. Such diseases will be correlated in population studies but not in case-control studies. In other examples the two are causally related with one being the precursor of the other. Such lesions will be associated epidemiologically both in population and in case-control studies because they are on a common pathway. Some known examples of associated lesions and precursor states are given in Table 1 for colorectal cancer. For a lesion to be classified as precancerous it must have been shown to be on the multistage histopathological pathway from normal mucosa to cancer. Thus, since all colorectal cancers arise in precursor areas of dysplasia, and since an adenoma is defined histologically as an area of dysplasia, adenomas are precursors of colorectal cancer. Since it is known that dysplasia arises at above normal frequency in inflamed and ulcerous tissue, pancolitis is also a precursor lesion. In contrast, cancer of the endometrium is strongly correlated with colorectal cancer but only because they both have a common correlation with intake of fat and calories. These cancers are

Table 1. Coincidental correlations, associated diseases and precursor lesions to colorectal cancer.

Status	Lesion	Underlying cause of correlation
Coincidental	hamartomatous polyps of the colon	none
Associated	haemorrhoids	}
	diverticular disease	} low fibre
	appendicitis	} diet?
	diabetes	}
	hormone-related	}
	cancers	} high fat diet?
	gallstones	}
	heart disease	}
Precursor lesions	adenoma	precursor
	polyposis coli	gentic predisposition to adenomas
	chronic pancolitis	precursor

```
                                        /-------> lesion B
Normal mucosa ----> lesion A -
                                        \-------> lesion C
```

Figure 1. The relation between associated and precursor lesions. A is the precurosr to B and C; the latter are associated and lesion B does not predispose to C.

therefore associated lesions. The inter-relationship between associated and precursor lesions is illustrated in Figure 1.

Table 2 lists some examples of precursor lesions to cancers at various sites. Clearly, much more is known about the histopathological sequence of events leading to accessible cancers (such as cancer of the digestive tract, the cervix or endometrium) than to deep tissue malignancies such as cancer of the breast, liver, pancreas, brain etc.

Table 2. Some examples of precursor lesions

Site	Precursor lesion	Carcinoma
Mouth	Leukoplakia	Squamous cell ca
Oesophagus	Achalasia	Squamous cell ca
	Barrett's oesophagus	Adenocarcinoma
Stomach	Pernicious anemia	Adenocarcinoma
	Intestinal metaplasia	
Colorectum	Adenoma	Adenocarcinoma
	Pancolitis	Adenocarcinoma
Cervix	CIN	Squamous cell ca

WHY STUDY PRECURSOR LESIONS?

The rationale for studying precancerous lesions includes both academic and practical considerations. Carcinogenesis often involves a multistage sequence of precancerous lesions, each step having its own set of causal agents (Figure 2). The absence of any of these will interrupt the process of carcinogenesis regardless of the amounts of the other factors, and this can result in a very confused epidemiology with only weak correlations being detectable. If the sequence of events is known then the causation of the individual stages can be studied in isolation leading to improved understanding of the causation of the overall disease process. At the practical level it also gives important clues to strategies for cancer prevention by the inhibition of one or more of the precursor stages. Because of the level of understanding that is possible with this route to cancer prevention the monitoring of the intervention is easier and so the safety of the strategy can be optimised. When a precancerous lesion is being treated then any treatment failure leaves the patient either still in a precancerous state (in which case an alternative strategy can be tried) or at worst with an early cancer which is usually amenable to surgical removal. In contrast, when the aim is to prevent cancer per se then treatment failure leaves the patient with full-blown cancer and a poor prognosis.

DIET AND PRECANCEROUS LESIONS

The most well-studied precancerous lesions will be discussed in detail by others in this symposium in the papers on gastric and colorectal cancer. The examples of precursor lesions that I will consider here are therefore ones which have been less well studied.

Oropharyngeal Cancer

Although this cancer is of relatively minor importance in the western countries (other than in the francophone European countries of France, Luxembourg and Switzerland) it is the third commonest cancer site in the developing world (particularly eastern Asia) and is therefore of great importance on the world scale. In the western countries, although the disease is not common, it is rapidly progressing and the histological sequence is not necessarily apparent. In contrast, in Asia the disease is relatively slow to evolve and then the precursor lesions are clearly identifiable. The incidence of the cancer is strongly correlated with the intake of alcohol and with the smoking of tobacco; it is also associated with poor nutrition and particularly with a low intake of retinoids and of carotenoids. The epidemiology of cancers of the mouth and pharynx has been reviewed recently by Johnson, 1991.

The commonest precursor lesion is leukoplakia, which has a prevalence of up to 5% in India but only 1% in the low-risk western countries. It has a 2-6% risk of progressing to squamous cell carcinoma through oral epithelial dysplasia (Johnson, 1993). In epidemiological studies its prevalence has been associated with low intakes of retinoids and carotenoids (Garewal, 1991) which suggests that the risk factors of the cancer are dominated by those of the precancerous lesion. The use of retinoid supplements has not been attempted widely because of the well-known toxicity problems but in preliminary studies the lesion has been shown to be reversed by dietary supplements of carotene (Garewal, 1991).

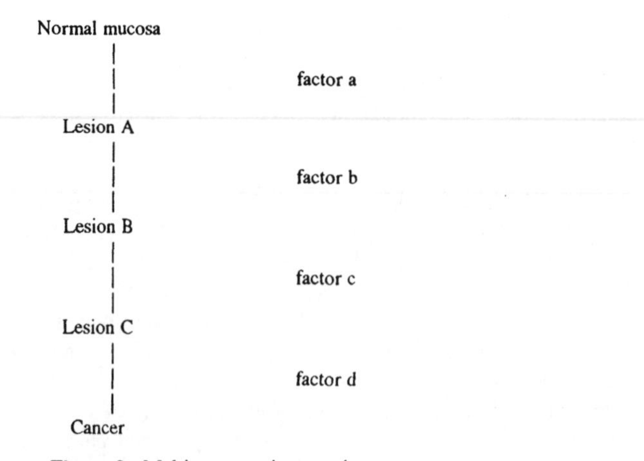

Figure 2. Multistage carcinogenesis.

Oesophageal Cancer

The world distribution of oesophageal cancer is similar to that of the oropharynx. In epidemiological studies the risk of the disease has been correlated with intake of alcohol and tobacco usage in the West, and with poor nutrition in the developing world. The major histological type of cancer is squamous cell carcinoma (SCC); in most reports SCC has accounted for more than 90% of all cases. The most widely recognized precursor of SCC is dysplasia in achalasia; achalasia carries a 33-fold excess risk of SCC (Meijssen et al., 1992) and there have been numerous hypotheses concerning its causation. In the large studies carried out in China the major risk factors for dysplasia were low intakes of retinol, riboflavin and zinc (Thurnham et al., 1982). In a follow-up dietary intervention study, however, supplements of these were not able to reverse the development of these lesions (Munoz et al., 1985). Others have suggested a causal role for tannins (as chelating agents for zinc) in the development of achalasia (Craddock, 1987).

In recent decades there have been reports from the USA and the UK of a rising trend in incidence of adenocarcinoma (AC) of the oesophagus, which now accounts for approximately 25% of all oesophageal cancers in many countries (Powell and McConkey, 1992; Tuyns, 1992). In a recent workshop organized by the European Cancer Prevention Organization (Reed, 1991) it emerged that AC was now a major contributor to the total risk of oesophageal cancer in many northern European countries and in areas of France and Switzerland. The most widely recognized precursor lesion is Barrett's oesophagus, which progresses through intestinal metaplasia then increasingly severe dysplasia and finally AC of the oesophagus (Miros et al., 1991; Bartelsman et al., 1992). There is, as yet, no information on the role of diet in the causation of Barrett's oesophagus; the only information is that alcohol and tobacco play no part in its causation (Levi et al., 1990). In consequence we have no clues to the prevention of this cancer even though the multistage nature of its histogenesis is known.

CONCLUSIONS

No disease has ever been controlled through treatment. There is considerable scope for study of precancerous lesions because of the leads that they can give to cancer prevention. In particular, the role of diet in their causation is of particular importance because although the role of overall diet in carcinogenesis at a number of sites is clear the associations with particular dietary items tend to be weak. In these situations the role of diet in individual steps in the multistage carcinogenesis process is likely to be much clearer and is likely to provide the best clues to strategies for cancer prevention.

REFERENCES

Bartelsman, J.R., Hameeteman, W., Tytgat, G.N., 1992, Barrett's oesophagus. Eur. J. Cancer Prev. 1:323.

Benito, E., Giacosa, A., Hill, M., 1992, "Public Education on Diet and Cancer," Kluwer, Lancaster. Carroll, K.K., 1985, Diet and breast cancer, in: "Diet and Human Carcinogenesis," J. Joossens, J. Geboers, M. Hill, eds., Excerpta Medica, Amsterdam.

Craddock, V.M., 1987, Nutritional approach to oesophageal cancer in Scotland, Lancet i:217.

De Waard, F., 1992, Diet and breast cancer, in: "Public Education on Diet and Cancer," E. Benito, A. Giacosa, M. Hill, eds., Kluwer, Lancaster.

Doll, R., Peto, R., 1981, The causes of cancer: quantitative estimates of available risks of cancer in the United States today, J.N.C.I. 66:1191.

Garewal, H.S., 1991, Potential role of B-carotene in prevention of oral cancer, Am. J. Clin. Nutr. 53:294S.
Johnson, N.W., 1991, Orofacial neoplasms: global epidemiology, risk factors and recommendations for research, Int. Dental J. 41:365.

Johnson, N.W., 1993, Histopathology of precancerous lesions of oral cavity and pharynx, in: "Precancerous Lesions of the Digestive Tract: Strategies for Intervention," A. Giacosa, M. Hill, eds., Eur. J. Cancer Prev. 2 (Supplement 2) in the press.

Levi, F., Ollya, J.B., La Vecchia, C., et al., 1990, The consumption of tobacco, alcohol and the risk of adenocarcinoma in Barrett's oesophagus, Int. J. Cancer 45:852.

Meijssen, M.A., Tilanus, H.W., van Blankenstein, M., et al., 1992, Achalasia complicated by oesophageal squamous cell carcinoma: a prospective study in 195 patients, Gut 33:155.

Miros, M., Kerlin, P., Walker, N., 1991, Only patients with dysplasia progress to adenocarcinoma in Barett's oesophagus, Gut 32:1441.

Munoz, N., Wahrendorf, J., Bang, L.J., et al., 1985, No effect of riboflavine retinol and zinc on prevalence of precancerous lesions of oesophagus, Lancet ii:111.

Powell, J., MacConkey, C.C., 1992, The rising trend in oesophageal adenocarcinoma and gastric cardia, Eur. J. Cancer Prev. 1:265.

Reed, P.I., 1991, Changing pattern of oesophageal cancer, Lancet ii:178.

Reed, P.I., 1992, Diet and gastric cancer, in: "Public Education on Diet and Cancer," E. Benito, A. Giacosa, M. Hill, eds., Kluwer, Lancaster.

Thurnham, D.I., Rathakette, P., Hambridge, K.M., et al., 1982, Riboflavin, vitamin A and zinc status in Chinese subjects in a high-risk area for oesophageal cancer in China, Hum. Nutr. 36:337.

Tuyns, A.J., 1992, Oesophageal cancer in France and Switzerland: recent time trends, Eur. J. Cancer Prev. 1:275.

Wynder, E.L., Gori, G., 1977, Contribution of the environment to cancer incidence: an epidemiologic exercise, J.N.C.I. 58:825.

DIETARY PREVENTION OF CHRONIC DISEASES: THE POTENTIAL FOR CARDIOVASCULAR DISEASES

Salvatore Panico[1], Egidio Celentano[2], Rocco Galasso[1], Eduardo Farinaro[3], Camilla Ambrosca[1], Rossano Dello Iacovo[2].

[1] Istituto Medicina Interna e Malattie Dismetaboliche
Facoltà di Medicina
Università Federico II
Napoli, Italy
[2] Direzione Scientifica
Istituto Nazionale Tumori di Napoli, Italy
[3] Cattedra Medicina di Comunità
Facoltà di Medicina Università Federico II
Napoli, Italy

INTRODUCTION

The evidence of the key role played by dietary habits on the frequency of the major killing chronic diseases, i.e. cancer and cardiovascular, indicates that the knowledge available on the relationship between diet and cancer needs to be interpreted in the light of data on cardiovascular disease. There are two main reasons for this: a) the possible common clues for the pathophysiology of cancerogenesis and atherosclerosis; b) the analysis of the possible benefits achievable in the general population through dietary preventive measures supposedly effective for both cancer and cardiovascular diseases.

This paper focuses on some "classical" dietary factors known to be related to cardiovascular diseases (fats, salt, and alcohol), trying to estimate the possible benefits of preventive measures based on these factors. The estimation is made with reference to the Italian population, in which the wide variety of food items consumed and the consequent wide variance of most nutrients allows efficient within-population contrasts.

THE IMPACT OF CARDIOVASCULAR DISEASES ON DEATH RATES

In all the economically developed countries cancer and cardiovascular diseases account for about two-thirds of all deaths (1). These diseases have also a great impact in terms of morbidity

Advances in Nutrition and Cancer, Edited by
V. Zappia *et al.*, Plenum Press, New York, 1993

and early disability, because of the great burden on medical expenditures at the individual and community levels

A simple overview in any such country of the mortality data on the most frequent cardiovascular diseases (coronary heart disease, stroke, hypertension) and cancers (breast, colon, and lung), reveals that in both men and women the events known to be influenced by dietary habits, account for a great number of deaths. This is also true when truncating the age range at 75 years, as is evident from the mortality data in Italy, reported as an example in Table 1 (2).

The data reported for diet-influenced diseases derive from all the deaths due to myocardial infarction, stroke, hypertension, breast cancer, and colo-rectal cancer, plus 50% of the deaths due to other ischemic heart diseases (trying to minimize certification bias) and 20% of the deaths from lung cancer (a reasonable percentage not attributable to smoking or occupational exposure). This can be considered a rough estimate of the number of deaths that are likely to have a relationship with dietary habits. These figures are not meant as deaths attributable to diet, but a reasonable estimation of those deaths where dietary habits might have played a role.

Table 1. Numbers and percentages of deaths due to diseases known to be influenced by diet; Italy, 1988.

	ALL	DIET-INFLUENCED	OTHERS
Men and Women (0-74 years)	71,966 (100%)	40,666 (56.5%)	36,559 (43.5%)

THE DIET-HEART HYPOTHESIS

The association between a "rich" diet and cardiovascular risk is the basis of the diet-heart hypothesis. This diet is defined as a habitual eating pattern characterized by high caloric intake compared to energy expenditure and a high content of cholesterol, saturated fat; it is usually also high in total fats, refined sugars, sodium and alcohol. It is a diet rich in animal products such as meats, eggs, dairies. It is in effect a diet with a high caloric density. During the last few decades with the modern development of the industrialization of human nutrition, this kind of diet has been largely adopted by several "western" populations.

This lifestyle has been seen as the key responsible determinant of the high frequency of cardiovascular diseases in those populations. The Seven Countries Study was the pioneer study designed by Ancel Keys to analyze the dietary and metabolic causes of the difference in cardiovascular risk between the Mediterranean countries (Italy, Greece, Yugoslavia) and some other countries including the USA, Finland, The Netherlands, and Japan. The cross-cultural analysis of these populations has supported - since the earliest papers - the relationship between dietary lipids and cardiovascular risk (3), in particular saturated fats and coronary risk. The same analysis has provided evidence for the association between saturated fats and serum cholesterol, and between serum cholesterol and coronary heart disease incidence.

The vast literature on cardiovascular diseases is consistent in defining the major risk factors, whose etiological significance is strongly supported by the consistency in several populations, the strength of the association, its graded nature (increasing risk with exposure to higher doses), and independence from other factors. High levels of serum cholesterol and blood pressure are the two among the major risk factors, that may be influenced by diet. For this reason the direct influence, i.e. not mediated by serum cholesterol and blood pressure, of dietary habits on cardiovascular risk has not been easily analyzed in epidemiological studies. Moreover, the dietary methods used have seldom provided sufficiently accurate information to identify individual dietary habits. Therefore only a few prospective studies have been able to discriminate the role of dietary factors independently from the other major risk factors. A few long-term studies have so far shown that dietary lipids influence not only serum cholesterol, but are also independently related to the risk of one of the major cardiovascular complications of atherosclerosis, i.e. fatal coronary heart disease (4-10). The populations examined by these studies are quite different from one another: American, Irish, Dutch, Japanese. The years of observation range from 10 to 25, and in all of them a quite accurate dietary method was the basis of the analysis (dietary history).

The relative risks are of a similar magnitude across the studies, and it is worthwhile to look at them through the results of the Western Electric Study in Chicago (4, 10). These results reveal that with dietary cholesterol intake lower by 200 mg/1,000 kcal (difference between 100 mg/1,000 kcal and 300 mg/1,000 kcal) and serum cholesterol lower by 40 mg/dl (difference between 220mg/dl and 180 mg/dl), the effects on the relative risk for coronary heart disease and all-cause mortality are multiplicative: for coronary heart disease, 0.53x0.80=0.42; for all-cause, 0.63x0.88=0.55. These values are adjusted for other major risk factors, including age, blood pressure, and cigarette smoking. Thus, mortality rates are lower by 58% and 45%, respectively. In the same study a similar analysis was carried out using the Dietary Lipid Score by Keys as a measure of consumption of saturated and polyunsaturated fats, and cholesterol (11). Comparing a 30-point difference in this score (reduction of saturated fats from 16% of total calories to 9%, of dietary cholesterol from 250mg/1,000 kcal to 100 mg/dl 1,000 kcal, and increase in polyunsaturated from 5% to 8%), the adjusted relative risk of coronary heart disease and all-cause mortality are respectively 0.45 and 0.62, while the synergistic action of these differences in dietary lipids and serum cholesterol (190 mg/dl vs 220 mg/dl) indicates relative risks of 0.38 and 0.56, with a decrease in death risk by 62% and 44%, which confirms the multiplicative effect. The greater longevity estimated for a man aged 50 coming from such a population is 3.8 years for coronary heart disease and 4.5 years for all-cause. If these analyses are applied to high risk individuals, the advantage in terms of life expectancy derived from a possible change in dietary habits and serum cholesterol increases considerably: 6.8 years for coronary heart disease and 8.3 for all-cause.

As pointed out earlier, the magnitude of the risk found in all these prospective studies is quite similar. In the Ireland-Boston Diet-Heart Study the relative risk -- adjusted for blood pressure, serum cholesterol, smoking, electrocardiographic abnormalities, and alcohol intake -- for a coronary event in individuals belonging to top tertile of the distribution of the Keys Dietary Lipid Score is 1.6 compared to those in the low tertile, meaning a 60% greater frequency of the disease in individuals who eat food containing more saturated fats and cholesterol, less polyunsaturated fats (8).

Recently the diet-heart hypothesis is developing in the direction of a possible vascular protection due to antioxidants, especially vitamins and micronutrients, of which vegetables and fresh fruits are rich. Plasma levels of vitamins C, E, and carotene have been found to be inversely associated to angina pectoris (12). In smokers the protection from carotene is cancelled, that one from vitamin C is reduced, while that from vitamin E persists, suggesting that an elevated consumption of vitamins E and C are important for the dietary prevention of coronary heart disease.

SALT, ALCOHOL, AND BLOOD PRESSURE

Blood pressure - both systolic and diastolic - is a major risk factor for cardiovascular disease and has a risk function that is very similar to serum cholesterol. Its major determinants are related to dietary habits such as excessive salt and alcohol consumption, excess weight, and dietary lipids.

The literature on these topics is quite extensive, however it is noteworthy to mention the results of the INTERSALT study, particularly for their relevance to the etiological role that salt and alcohol play in the development of high blood pressure levels and hypertension (13). The cross-cultural analysis carried out in 52 centers in 15 countries has shown the association between sodium excretion in the urine and the level of both systolic and diastolic blood pressure. This association persists even when the interaction with age is taken into account, revealing that whatever the cause of sodium excretion, this is the strongest determinant of blood pressure levels. Sodium excretion at a population level is a marker of dietary sodium intake, which is virtually salt consumption. INTERSALT has also shown that alcohol consumption and excess weight determine the population differences in both systolic and diastolic blood pressure. For dietary lipids there is increasing evidence of their influence on blood pressure regulation (14).

DIETARY HABITS AND CARDIOVASCULAR RISK IN THE ITALIAN POPULATION

The epidemiological studies carried out in Italy in recent decades have confirmed that the classical major risk factors play an etiological role also in the Italian population (15-21). The linearity of the risk due to these factors is also confirmed in these studies.

Little information relating dietary habits to cardiovascular incidence and mortality is available in Italian population groups: it is virtually confined to the Italian component of the Seven Countries Study (22). In this study none of the analyses presented was able to discriminate the specific influence of the single nutrients, since the individuals were classified according to the pattern of nutrients' aggregation. According to this original analysis, the individuals at high risk of coronary heart disease were found among those who were classified in a group characterized by the highest intake of alcohol and the lowest of carbohydrate. It should be recalled that the data pool from all the countries participating to the study had confirmed, on an ecological basis, that dietary fats were determinants of coronary heart disease (3).

There is very valuable information available on the relationship between dietary habits and major cardiovascular risk factors. An important source of data is the Nine Community Study carried out at the end of the seventies in random samples of the Italian population (23). According to these data the Italians habitually consuming high quantities of food items rich in animal fats have plasma cholesterol, triglyceride, glycemia and systolic and diastolic blood pressure higher than those consuming less (24). It has been found that the use of monounsaturated fats (due to the consumption of olive oil) contributes to lower levels of plasma cholesterol and blood pressure (25). The Italian component of the INTERSALT Study has confirmed that salt intake, alcohol consumption and excess weight influence blood pressure levels also in the Italian population (26).

In a group of families living in Southern Italy -- where the Mediterranean diet is the common way of eating -- a dietary experiment was carried out to study the influence of dietary changes on plasma cholesterol and blood pressure (27). Changing the diet for six weeks, increasing the intake of animal fats and reducing the polyunsaturated/saturated (P/S) ratio, plasma cholesterol and blood pressure raised significantly, going down after resuming the original diet. The inverse was observed in a specular experiment in Finnish families: decreasing animal fat consumption and increasing the P/S ratio lead to reduction in plasma cholesterol and blood pressure (28).

THE PREVENTIVE POTENTIAL OF THE REDUCTION IN FATS, SALT INTAKE, AND EXCESSIVE ALCOHOL CONSUMPTION IN ITALY

On the basis of the information published on samples of the Italian population, an estimation can be attempted to find out how many deaths due to total cardiovascular diseases, coronary heart disease, and stroke might be prevented in 25 years modifying the intake of atherogenic food, salt, and alcohol. Three major studies can be used for this analysis: the Italian Section of the Seven Countries Study, the Nine Communities Study of the National Research Council, and the Italian Section of the INTERSALT (23,24,26,29,30). Since the basis for the analysis is a male population sample, the estimation is applicable only to men (29).

There are data on the 25-year estimated probability of death from total cardiovascular diseases, coronary heart disease and stroke from a sample of 1,530 men aged 40-59 years at entry in the Italian section of the Seven Countries Study. The sample was first seen in 1960, and all the major risk factors have been measured since then. For this sample the multiple logistic function was calculated and the coefficients to solve the equation for total cardiovascular disease, coronary heart disease, and stroke were reported (29).

Data on the distribution of cardiovascular risk factors are available from nine randomized samples of the Italian population aged 20-59 years, with an even geographical representation for a total of more than 7,000 individuals of both sexes. Highly standardized procedures were adopted for biological measurements, and a food frequency questionnaire on dietary habits was available for all the participants (23). Among the many papers in relation to this study, there is one focusing on the influence of atherogenic food items on serum cholesterol and blood pressure (24). According to these results, there is a linear trend in the association between the intake of atherogenic food items, indicated through a dietary atherogenic index score (DAI), and both plasma cholesterol and blood pressure in the two sexes. The analysis was made by tertiles of the distribution of DAI and the mean level of plasma cholesterol and systolic blood pressure in the three tertiles were calculated adjusting for age, body mass index, and use of fats in the diet.

The four Italian INTERSALT centers were distributed across the country and constituted about 800 individuals aged 20-59 years of both sexes (26). The specific contributions of the level of sodium excretion (as a group indicator of salt intake), and alcohol consumption on systolic blood pressure were available separately and depurated from the specific effects of potassium excretion, body mass index, and age. No sex interaction was found, therefore the findings are applicable to both men and women. The data on the relationship between alcohol consumption and blood pressure could also be used as a cross check (28).

The number of expected deaths per 1,000 Italian men in 25 years for total cardiovascular diseases, coronary heart disease, and stroke have been calculated from the solution of the multiple logistic functions to predict total cardiovascular, coronary and cerebrovascular deaths in individuals with a number of dietary characteristics: low consumption of atherogenic food (low tertile of DAI distribution), high consumption of atherogenic food (top DAI tertile), daily consumption of 10 gr, 7 gr, and 4 gr of sodium chloride, and heavy, moderate and no alcohol consumption.

As for total cardiovascular deaths, the solution of the multiple logistic function indicates a prediction of 71.16 deaths per 1,000 for the average individual of the top DAI tertile compared with 58.34 per 1,000 for the bottom DAI tertile. The deaths predicted for the average individual consuming 10 gr of NaCl per day are 43.62 per 1,000, compared to 41.94 per 1,000 for 7gr, and 40.26 for 4 gr. The predicted deaths per 1,000 for the average alcohol abstainer are 64.58 compared to 65.56 per 1,000 for the moderate drinker and 72.29 for the heavy drinker.

As for coronary heart disease, the predicted deaths in the top DAI tertile are 48.40 per 1,000, while in the bottom DAI tertile they are 38.69 per 1,000. 28.11 deaths per 1,000 are predicted for the 10 gr consumption, 27.03 per 1,000 for 7 gr, 25.95 per 1,000 for 4 gr of sodium chloride. 43.40 deaths per 1,000 are predicted for abstainers, 44.14 for moderate and 49.21 for heavy drinkers.

As for stroke 22.93, per 1,000 are the deaths in the top DAI tertile compared to 20.35 per 1,000 in the bottom one. 15.94 deaths per 1,000 are predicted for a 10 gr consumption, 15.33 per 1,000 for a 7 gr, 14.71 per 1,000 for a 4 gr consumption. 21.50 deaths per 1,000 are predicted for abstainers, 21.75 for moderate drinkers and 23.46 for heavy drinkers.

The total amount of avoidable deaths due to changes from high to low consumption of atherogenic food, and from 10 gr per day of sodium chloride to 7 gr per day, and from heavy to moderate alcohol consumption is 21.23 per 1,000 for total cardiovascular, 15.86 per 1,000 for coronary heart disease, 4.8 per 1,000 for stroke (Table 2). This prediction for the following 25 years applies to male individuals aged 20-59, living in Italy.

Table 2. Potentially avoidable deaths per 1,000 in a 25-year period, changing specific dietary habits in the Italian male population aged 20-59.

DIETARY CHANGES	TOTAL CVD	CHD	STROKE
High to low atherogenic food	12.82	9.71	2.58
10gr to 7 gr of NaCl	1.68	1.08	0.61
heavy to moderate alcohol	6.73	5.07	1.71
TOTAL	21.23	15.86	4.80

The procedure adopted for this estimation requires some comments. The multiple logistic function is derived from a population sample aged 40-59, while the application was made in the age range 20-59. It should be noted, however that all the functions were solved with the age variable always above 40 years, since this is the level of the mean age of the categories analyzed in the two studies (Nine Communities and INTERSALT-Italy); moreover, the accuracy of the estimation for the categories of individuals in the different tertiles of the distribution of DAI or for the different consumption of NaCl is quite acceptable.

This estimation is partial, since other possible dietary influences on cardiovascular deaths have not been taken into account. It may well be an understatement of the possible beneficial effects of the dietary intakes rich in vegetables and fruit that are typical of the Mediterranean diet - with a low consumption of atherogenic foods - since no information is available on the protective role of antioxidants both on the major risk factors and directly on morbidity and mortality. Moreover, the possible multiplicative effect of several protective factors is not included in this analysis, again causing an underestimation of this protection. The generalization of these data to the Italian population is based on the fact that the basic information derived mostly from representative population samples.

This estimation gives some interesting suggestions for the prevention of cardiovascular risk through risk pattern modification in population, since it derives from a population consuming a wide variety of food items; this allows a within-population observational analysis which is hardly available elsewhere in Europe. The consumption of atherogenic food items varies widely across the country, and the difference between the continental and the Mediterranean part of the country is quite high. The DAI used in the Nine Communities Study as an indicator of atherogenic food consumption derives from a food frequency questionnaire. The score has been computed on the basis of the known amount of total and saturated fat of the different food items consumed. The tertiles of distribution of this score largely represented the contribution of the different centres participating in the Study. The Mediterranean Centers contribute mostly to the bottom tertile while the continental parts largely contribute to the top tertile. This is important in terms of feasibility of the preventive action, since the availability of all the principal food items involved in determining the DAI score is virtually ubiquitous all over the country. The safety of having a low DAI score is indirectly supported by the low incidence and mortality data for all the major chronic diseases (cardiovascular and cancer). In the Mediterranean part of Italy incidence and mortality rates for cancer are about half of those of the Continental part, and this is only partly due to the rate of industrialization, since cancers that are only slightly related to the effect of a highly industrialized environment (breast and colon) show the highest geographical differential; moreover, there are some indications that dietary habits may be the major cause for this difference. In addition the change to an intake of 7 gr of sodium chloride per day and to a moderate alcohol consumption is quite feasible and does not have any undesirable effects on health, so that the protective dietary changes we have analyzed are safely advisable.

It is important to note that these changes may be useful also in the dietary prevention of cancer, and that the number of total deaths avoidable when these are applied is certainly higher and of great interest for prevention programs in the community.

It should be emphasized that all the estimations attempted in this paper are made on men, since there is no information on the relationship between diet and incidence or mortality of cardiovascular diseases in Italian women. The body of knowledge on cardiovascular risk factors is very recent in the international scientific literature, because of the lower frequency of the disease - particularly coronary heart disease in pre-menopausal women - and for the consequent difficulties to carry out epidemiological investigations on large numbers of individuals. Recently, studies of a new generation have been started: these are prospective investigations based on a biological specimen bank, in order to increase the efficiency and reduce the costs. Among these is a new on-going project in the city of Naples, named Progetto ATENA, aimed to study the etiology of major chronic diseases in women, i.e. cardiovascular and the most frequent cancers, with special attention to the role of dietary habits (31). The interest of this type of investigation is related to both etiopathogenesis and prevention, and may provide comprehensive information in the near future.

ACKNOWLEDGMENTS

This paper has been prepared in the framework of the activities of the PROGETTO ATENA, funded by National Research Council: Targeted Project "Prevention and Control Disease Factors"; Subproject "Community Medicine"; grant No.92 00149.PF41. The authors thank Ms Rosanna Scala for the linguistic revision.

REFERENCES

1. WHO Statistics (1987).
2. ISTAT, Statistiche di Mortalità (1988).
3. A. Keys. "Seven Countries Study. A Multivariate Analysis of Death and Coronary Heart Disease in 10 Years," Harvard University Press, Cambridge Ma USA, (1980).
4. J. Stamler, R.B. Shekelle, Dietary cholesterol and human coronary heart disease: the epidemiological evidence, Arch Pathol Lab Med. 112:1032 (1988).
5. R.B. Shekelle, A.M. Shryock, O. Paul, et al, Diet, serum cholesterol, and death from coronary heart disease: The Western Electric Study, N Engl J Med. 304:65 (1981).

6. R.B.Shekelle, O. Paul, A.M. Shryock, et al, Dietary lipids, serum cholesterol concentration, and risk from coronary heart disease, in: "Lipoproteins and coronary atherosclerosis," G.Noseda, et al, eds., Elsevier, Amsterdam (1982).
7. R.B. Shekelle, J. Stamler, Dietary cholesterol and ischemic heart disease, Lancet. i:1177 (1989).
8. L.H. Kushi, R.A. Lew, F.J. Stare, et al, Diet and 20-year mortality from coronary heart disease: The Ireland-Boston Diet-Heart Study, N Engl J Med. 312:811 (1985).
9. D.L. McGee, D.M. Reed, K. Yano, A. Kagan, J. Tillotson, Ten-year incidence of coronary heart disease in the Honolulu Heart Program. Am J Epidemiol. 119:733 (1984).
10. D. Kromhout, E.B. Bosschieter, C.L. Coulander, The inverse relation between fish consumption and 20-year mortality from coronary heart disease, N Engl J Med. 312:1205 (1985).
11. A. Keys, J.T. Anderson, F. Grande, Serum cholesterol response to changes in the diet. Metabolism. 14:747 (1965).
12. R.A. Riemersma, D.A. Wood, C.C.G. MacIntyre, R.A. Elton, K.F. Gey, M.F. Oliver, Risk of angina pectoris and plasma concentrations of vitamins A, C, and E, and carotene. Lancet. 337:1 (1991).
13. INTERSALT Cooperative Research Group, INTERSALT, an international study of electrolyte excretion and blood pressure. Results for 24-hour urinary sodium and potassium excretion, Br Med J. 297:319 (1988).
14. P. Pietinen, J.K. Huttunen, Dietary fat and blood pressure-a-review, Eur Heart J. 8(Supp.B):9 (1987).
15. Italian Research Group of the Seven Countries Study, 25-year incidence and prediction of coronary heart disease in two population samples, Acta Cardiol. 41:283 (1986).
16. A. Menotti, S. Conti, F. Dima, S. Giampaoli, B. Giuli, M. Matano, F. Seccareccia, Incidence of coronary heart disease in two generations of men exposed to different levels of risk factors, Acta Cardiol. 40:307 (1985).
17. A. Menotti, S. Conti, S. Giampaoli, S. Mariotti, P. Signoretti, Coronary risk factor predicting coronary and other causes of deaths in 15 years, Acta Cardiol. 35:107 (1980).
18. A. Menotti, G. Farchi, F. Seccareccia, R. Capocaccia, S. Conti, La predizione a breve termine degli eventi coronarici nel Progetto Romano di Prevenzione della Cardiopatia Coronarica, Clin Ter Cardiov. 2:193 (1983).
19. G.C. Descovich, G. Dalmonte, A. Dormi, A. Braiato, G. Mannino, M.N. Benassi, A. Gaddi, Z. Sangiorgi, N. Trivelli, S. Lenzi, The Brisighella Study: A community survey, in: "Atherosclerosis and Cardiovascular Disease," S. Lenzi, and G.C. Descovich, eds., MTP Press, Hingam USA. (1984).
20. G.C. Descovich, A. Dormi, A. Gaddi, G. L. Magri, G. Mannino, S. Rimondi, P. Perini, Z. Sangiorgi, S. Lenzi, The Brisighella Study, in: "Atherosclerosis and Cardiovascular Disease," S. Lenzi, and G.C. Descovich, eds., MTP Press, Norwell MA USA. (1987).
21. A. Spagnolo, A. Menotti, S. Giampaoli, G. Morisi, A. Buongiorno, G.C. Urbinati, G. Righetti, G. Ricci, High density lipoprotein cholesterol distribution and its predictive power in some Italian population studies, Eur J Epidemiol. 5:328 (1989).
22. G. Farchi, S. Mariotti, A. Menotti, F. Seccareccia, S. Torsello, F. Fidanza, Diet and 20-y mortality in two rural population groups of middle-aged men in Italy, Am J Clin Nutr. 50:1095 (1989).
23. The Research Group ATS-RF2 of the Italian National Research Council, Distribution of some risk factors for atherosclerosis in nine Italian population samples, Am J Epidemiol. 113:338 (1981).
24. M. Trevisan, V. Krogh, J. Freudenheim, A. Blake, P. Muti, S. Panico, E. Farinaro, M. Mancini, A. Menotti, G. Ricci, Diet and coronary heart disease risk factor in population with varied intake, Prev Med. 19:231 (1990).
25. M. Trevisan, V. Krogh, J. Freudenheim, A. Blake, P. Muti, S. Panico, E. Farinaro, M. Mancini, A. Menotti, G. Ricci, Consumption of olive oil, butter and vegetable oils and coronary heart disease risk factors, JAMA. 263:688 (1990).
26. E. Farinaro, M. Trevisan, F. Jossa, S. Panico, E. Celentano, M. Mancini, S. Zamboni, C. Dal Palù, F. Angelico, M. Del Ben, M. Laurenzi, INTERSALT in Italy: Findings and community health implications, J Hum Hypert. 5:15 (1991).

27. A. Ferro-Luzzi, P. Strazzullo, C. Scaccini, A. Siani, S. Sette, M.A. Mariani, P. Mastranzo, R.M. Dougherty, J.T. Jacono, M. Mancini, Changing the Mediterranean diet: effect on blood lipids, Am J Clin Nutr. 40:1027 (1984).
28. C. Enholm, J.K. Huttunen, P. Pietinen, et al, Effect of diet on serum lipoprotein in a population with a high risk of coronary heart disease, N Engl J Med. 307:850 (1982).
29. A. Menotti, F.Seccareccia, Blood pressure, serum cholesterol and smoking habits predicting different manifestations of arteriosclerotic cardiovascular diseases, Acta Cardiol 2:91 (1987).
30. M.Trevisan, V.Krogh, E.Farinaro, S.Panico, M.Mancini, Alcohol consumption, drinking pattern, and blood pressure: analysis of the data from the Italian National Research Council Study, Int J Epidemiol, 16:520 (1987).
31. S. Panico, R. Dello Iacovo, E. Celentano, R. Galasso, P. Muti, M. Salvatore, M. Mancini, PROGETTO ATENA, A Study on the Etiology of Major Chronic Diseases in Women: Design, Rationale and Objectives, Eur J Epidemiol, 4:601 (1992).

DIETARY FIBRES AND CANCER

Attilio Giacosa [1], Rosangela Filiberti [2], PaolaVisconti [1], Michael J.Hill [3], Franco Berrino[4], and Amleto D'Amicis[5]

[1] Unit of Clinical Nutrition, National Institute for Cancer Research, Genova, Italy
[2] Department of Epidemiology and Biostatistics, National Institute for Cancer Research, Genova, Italy
[3] ECP (UK), PO Box 1199, Andover, Hants SP10 1YN, U.K.
[4] Epidemiology Unit, National Cancer Institute, Milano, Italy
[5] National Institute of Nutrition, Rome, Italy

INTRODUCTION

by A. Giacosa, R. Filiberti, and P. Visconti

The concept that dietary fibres play a relevant role in human health, comes from the observation that some of the most frequent diseases in western and more industrialized countries show a very low incidence in developing areas. Dietary habits in these two areas are very different since the former area is characterized by a diet more rich in fat, calories and refined foods, than the developing countries.

The hypothesis that a low fibre intake may play a role in etiology of some chronic-degenerative diseases leads us to examine thoroughly the nature of fibres, together with their biologic and functional characteristics.

Dietary fibres are a heterogeneous entity of vegetable components deriving from different parts of the plant and may have different physiological effects. A clear definition of dietary fibre is complex, since various methods of analysis exist. Chemico-physical properties and function of fibres change according to the variation of compound percentage and water solubility.

Nowadays, the term "dietary fibre" is unsatisfactory for many reasons and a new classification of plant polysaccharides has been adopted on the basis of digestibility and chemical composition. This new classification divides plant polysaccharides into Starch Polysaccharides (SP) and Non-Starch Polysaccharides (NSP) (Englyst et al.,1990): the former ones are hydrolised by pancreatic α-amylase, while the latter are completely resistant to digestion by the enzymes of the human small bowel (Leeds, 1991).

Advances in Nutrition and Cancer, Edited by
V. Zappia *et al.*, Plenum Press, New York, 1993

According to these considerations, NSP could be identified as "dietary fibres", but recently Englyst and his collaborators (1990) showed that a part of NSP is degraded in the small intestine in some animals; while some starch (called Resistant Starch (RS)), passes into the colon, thus behaving as "fibre". Among RS, a great variability exists due to the cooking methods and food preservation (Leeds,1991).

From a practical point of view, this new classification implies difficulties in defining correctly the true type of fibre intake. It may lead to clinical implications, accounting, at least in part, for the literature discrepancies on the correlation between dietary fibre and some diseases, in particular colon cancer (Leeds, 1991).

Urbanization and industrialization in western countries are two factors linked to a decreased dietary fibre intake, with an increase of foods rich in energy. This unbalanced way of eating favours the increase of the so called "civilization diseases" such as cardiovascular diseases, diabetes, obesity, gastroenterological diseases and some site specific tumours, like colon and breast cancer.

In spite of the great interest in the relationship between dietary fibre and human health, data on the preventive role of specific fibre components are still controversial. The difficulties in assessing individual intake of dietary fibre must be taken into account to explain the inconsistencies of some of these findings. Moreover, the great variability of the metabolic effectiveness of different fibre types and the presence of other dietary components (which may interact as protective or risk factors in the etiogenesis of various diseases) have to be considered.

Dietary fibre is supposed to improve tolerance to glucose and decrease concentrations of atherogenic lipoproteins (Rivellese et al.,1980; Rivellese et al., 1983). Evidence demonstrating the effect of dietary fibre on blood lipids has accumulated mainly over the last fifteen years. Generally speaking, fibre decreases the amount of triglycerides and of cholesterol, which are absorbed during the food transit in the G.I. tract. This reduced absorption lowers the number of lipid molecules that can enter the blood and lymphatic stream.

Hypertension is related to atherosclerosis and CHD, and blood pressure is supposed to be influenced by dietary fibre, together with various other dietary factors (Mancini et al.,1983). There is not a good evidence that these factors may have a therapeutic benefit, but the complex carbohydrates (with its associated fibre) would be an important element for a preventive energy-reduced and low energy density diet.

As far as gastrointestinal diseases are concerned, dietary fibre may play a role since they influence the stool bulking and consistency. In addition to the prevention of constipation, fibre could help in preventing other intestinal pathologies such as diverticular diseases.

The revival in interest in dietary fibre has led to the suggestion that a high fibre diet would be beneficial as part of a low calorie regimen. Although there is no evidence that fibres *per se* have a direct effect in promoting satiety, high fibre diets seem more likely to produce appropriate sensation of satiety because the greater food bulk leads to more chewing and to a slower passage to the stomach (Eastowood and Passmore, 1983).

DIETARY FIBRE : THE ITALIAN SCENE

by A. D'Amicis

In the last decade, in Italy as in other countries, the interest in dietary fibre has greatly increased due to its possible protective role against many diseases, most of them affecting the gastrointestinal apparatus.

In Italy, data on dietary fibre consumption are rather scanty. However, it is possible to estimate the dietary fibre intake by means of food consumption data measured in several Italian food surveys and by analysis of Italian reference diet.

The sources of data representative at the national level are the Food Balance Sheets and the Household Budget Surveys both produced by the National Institute of Statistics (ISTAT). Another source of data is the National Food Consumption Study conducted by the National Institue of Nutrition in the quinquennium 1980-1984.

The calculation of dietary fibre in foods is based on the Food Composition Tables of the National Institute of Nutrition (1989) with the addition of several unpublished data of the Food Chemistry Laboratory.

During the past ten years (1981-1990), a study on the trend of dietary fibre consumption (Turrini et al., 1992) has been carried out utilizing data of the Nationwide Households Food Consumption Surveys carried out by ISTAT (1981): data are collected every year on 36,000 households by the Household budget survey. Figures relative to 1981 and 1990 are reported in Table 1. As in some other European countries, dietary fibre consumption seems to have increased also in Italy in the recent decade. As shown in Table 1, an increase of about 5 g/day/person (about 20%) appears from 1981 to 1990, mostly due to the high increase of fruit and vegetables consumption.

Table 1. Changes in dietary fibre intake in Italy from 1981 to 1990 (g/day) estimated from nationwide household food consumption. ISTAT.

	1981	1990
Total dietary fibre	20.9	25.0
Cereals & derivates	11.7	12.1
Vegetables, fruits & legumes	9.2	12.9

(Turrini et al. 1992)

Previous data on dietary fibre content, as calculated from data of the Italian Cohort in the Seven Countries Study in the 1960s (Fidanza and Fidanza Alberti, 1971), show an average intake of about 24 g/day/person (ranging from 22 to 25 g/day/person) (Table 2).

Most likely a decrease in dietary fibre consumption occurred around the 1970s, since the 1960s' figure is slightly higher with respect to that derived from ISTAT data on 1981 and very close to those on the '90s.

The INN Nationwide Food Consumption Survey (1980-1984) was conducted having as its objective the evaluation of the real food intake (Saba et al., 1990); it was carried out by the weighted and inventory method for seven days on about 10,000 households, for a

Table 2. Dietary fibre intake in the 1960s estimated* from food consumption of Italian Cohorts in the Seven Countries Study (a).

	g/day
Montegiorgio	25.8
Crevalcore	24.9
Roma	21.7

* By means of factors calculated from Nationwide Food Consumption Survey INN (Saba et al. 1990).
(a) Fidanza and Fidanza Alberti (1971)

total of about 30,000 individuals, randomly selected from the electoral lists of 9 (out of 20) Italian regions from North to South, and represents the largest food consumption data-bank in Italy. Although based on the household as a sampling unit, this survey allowed the collection of sufficient information to evaluate the average level of individual intake. In Table 3 the results of dietary fibre consumption are presented as g/day/consumption unit (20.7 g). Figure 1 reports the source of dietary fibre from food groups and the subdivision in soluble and insoluble fraction; about 64% of total dietary fibre is represented by unsoluble fraction.

From the same data-bank, a simulated diet (Turrini et al., 1991) has been formulated as "raw diet" (that is, a mixture of food as bought) and as "cooked diet" (that is, a mixture of food as consumed); food items were also collected in 11 groups. Determination of dietary

Table 3. Dietary fibre intake in Italy (g/day) evaluated from nationwide food survey data (INN 1982-84).

	pro capite (a)	per unit of consumption (b)	analysis of simulated* diet (c)		calculated by simulated* diet (d)
Total dietary fibre	20.9	24.7	Raw	25.3	21.6
			Cooked	25.6	
Cereals & derivates		12.1		10.3	10.6
Vegetables		5.9		7.6	4.5
Fruits		4.9		6.1	4.9
Legumes		1.8		2.0	1.6

(a) Saba et al (1990). (b) Turrini et al (1992). (c) Cappelloni M. et al. personal communication, by AOAC method. (d) Cappelloni M. et al. personal communication, by Food Composition Tables INN (1992).
* Simulated diet from INN's Nationwide Food Consumption Survey data-bank.

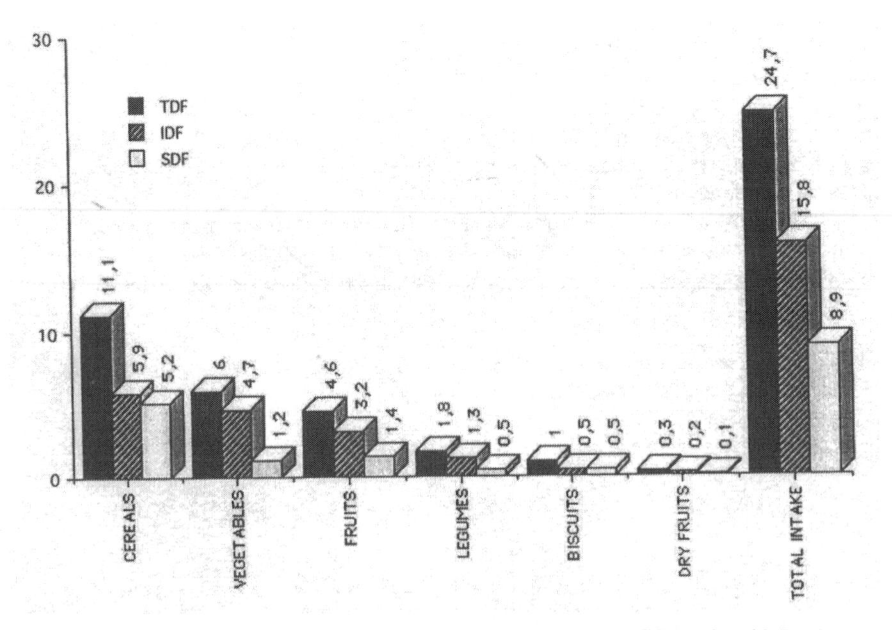

Figure 1. Dietary fibre intake in Italy (g/day/consumption unit). (Source: INN Nationwide Food Survey).

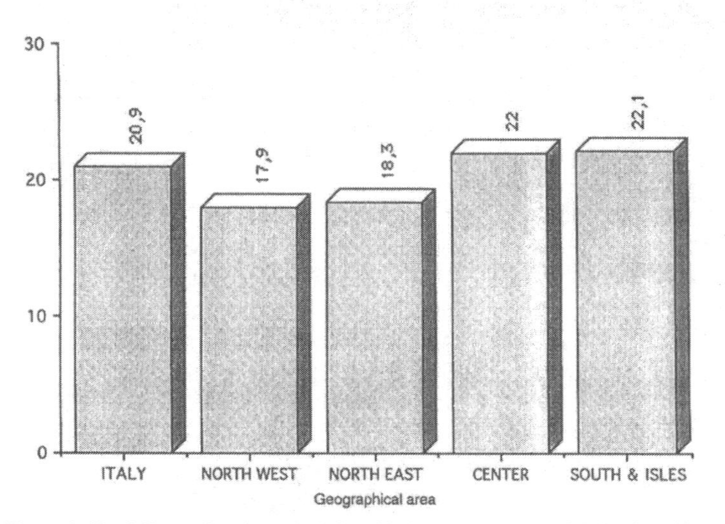

Figure 2. Total dietary fibre intake in Italy (g/day/person). (Source: INN Nationwide Food Survey).

Table 4. Dietary fibre intake in the 1960s calculated* from food consumption in two Italian areas (EURATOM Study)(a).

	g/day/consumption unit
North Italy	18.2
South Italy	24.4

* Calculated by Ferro-Luzzi and Sette 1989. (a) Cresta et al. 1969

fibre content of diet and plant foods by the AOAC method (1990) has been carried out by Cappelloni and Lintas, and some preliminary results are shown in Table 3. The intake of dietary fibre is 25 g/day, 40% of which is represented by cereals, 29% by vegetables, 23% by fruits, and 7% by legumes. The greatest differences between analyzed and calculated data make references to vegetables; there may be a double reason for this: the shortage of analytical data (so that the calculated data are underestimated), and the high variability of dietary fibre content of vegetables in comparison with other food groups.

At the geographical level (Fig. 2) the dietary fibre consumption, in terms of g/day/person, is higher in the Center and South Italy (Cresta et al., 1969), with respect to the North West and North East, of about 4 g/day/person (+5%).

These data confirm Ferro-Luzzi et al. (1989) who calculated fibre intakes from the EURATOM Food Consumption Survey, carried out during the early '60s in some European countries including North and South Italy. As is shown in Table 4, the consumption is higher in the South by about 6.2 g/day/consumption unit.

From the data presented it is possible to evaluate a mean dietary fibre intake in Italy of about 25 g/day. More research is necessary to better quantify the fibre consumption in our country; certainly, however, in the diet of the Italians, an additional 10 g of dietary fibre a day might find room to replace fats and energy.

FIBRE AND COLON CANCER: EPIDEMIOLOGICAL DATA

by R. Filiberti, P.Visconti, and A.Giacosa

Interest in the relationship between the fibre intake and colon cancer derives from Burkitt's observation of a low frequency of colon cancer in areas (such as Africa) characterized by a high fibre consumption and a high stool bulk (Burkitt, 1971; Willett et al.,1990). It has been shown that countries with a low consumption of cereals present high rates of colon cancer.

Moreover, an inverse association between overall fibre intake (and more precisely, from fruits or vegetables) and colon cancer risk has been shown by some case-control studies (Willett,1989). Nevertheless, correlations between fibre intake and cancer rates are low and findings from different epidemiological studies are controversial.

Early epidemiological studies on fibre intake and colon cancer were carried out in the 1970s. Many case-control studies supported a protective role of dietary fibre: among these, Modan et al. (1975) reported a lower consumption of high-fibre foods for colon cancer cases than for controls, while no difference was seen for rectal cancer.

In a study on a black population of Washington, Dales et al. (1978) showed that the risk of colorectal cancer was significantly increased with a high-fat/low fibre diet in comparison to a low-fat/high-fibre consumption. A protective effect was associated only with vegetable fibre intake and not with cereal fibre (Freudenheim and Graham,1989; Slattery et al.,1988; Willett et al.,1990). Bidoli et al (1992) observed a protective effect of whole grain bread and pasta against colorectal cancer in a case-control study in north-east Italy. In a case-control study on Japanese citizens emigrated to the US, dietary fibre showed a protective effect only when associated with a low fat intake. Macquart-Moulin et al. (1986) found that subjects consuming more fibre from vegetables presented a significant decrease in cancer risk, but the possible effects could not be distinguished from those of other vegetable components.

Other epidemiological studies failed to find any effect and in some surveys an increased risk associated with a high fibre intake has also been shown: for example Potter and MacMichael (1986) observed a positive correlation with fibre intake in one Australian study (Table 5) (Faivre,1991).

The main advantage of case-control studies is that non-dietary variables can be taken into account and this should make the results very much more reliable. However, in the case of colorectal cancer, where the onset of symptoms is insidious, a major disavantage of case-control studies is that recall methods have to be used to determine the diet before the onset of symptoms, since they affect the current diet. Such methods may be inaccurate and it is not surprising that the results from such studies have been highly variable.

The contradictory results shown in Table 6 on the protective effect of dietary fibre in colorectal cancer may be due also to the difficulties in assessing the individual intake of

Table 6 - The effect of some fibre sources on colonic physiology

	Stool bulk	Caecal pH	FBA concentration	Caecal gas production
Wheat bran	Increase+++	---	Decrease+++	+
Oat bran	Increase+++	--	No change	+
Ispaghula	Increase+++	Acid++	Decrease+++	+
Pectin	--	Acid+	Increase+	++
Guar	--	--	Increase+	++
Lactulose	Increase	Acid+++	Decrease+++	+++

Source: Wilpart,1987

Table 5. Association among fibre, fibre-providing foods, and risk of colorectal cancer from some case-control studies.

Study Area (Reference)	N° of cases	Fibre	Vegetables	Cereal Starch
Israel (Modan et al, 1975)	275	Decreased risk	--	--
USA (Graham et al, 1978)	586	--	Decreased risk	--
USA (Dales et al,1979)	99	(Decreased risk)	--	--
Puerto Rico (Martinez at al,1981)	461	Increased risk	--	--
Canada (Jain et al,1980; Miller et al,1983)	542	N.A.	--	--
UK (Bristol et al,1985)	50	N.A.	--	Decreased risk
France (Macquart-Moulin et al,1986)	399	Decreased risk	Decreased risk	--
Australia (Potter & McMichael, 1986)	419	Increased risk	--	--
Australia (Kune et al,1987)	715	Decreased risk	Decreased risk	Decreased risk
USA (Lyon et al,1987)	246	N.A.	--	--
Belgium (Tuyns et al,1987, 1988)	1235	Decreased risk	Decreased risk	Decreased risk
USA (Slattery et al,1988)	231	Decreased risk	--	--
USA (Young & Wolf,1988)	353	--	Decreased risk	--
USA (Graham et al,1988)	428	Decreased risk	--	--
Spain (Benito et al,1991)	286	Decreased risk (from legumes)	--	--
USA (Freudenheim et al,1990)	422	Decreased risk (from vegetables)	--	--
Chinese in Singapore (Lee et al,1989)	203	--	Decreased risk	--
American Japanese (Heilbrun et al,1989)	162	Decreased risk (in low fat intake)	--	--
Italy (La Vecchia et al,1988)	575	--	Decreased risk	Increased risk

different types of dietary fibre. As a matter of fact, there is evidence that not all components of dietary fibres have the same effect and that they should not be considered as a whole, but that attention should be focused on different types of fibre.

It has also been suggested that an imbalance between fat and fibre intake may be a major determinant of colon cancer risk, greater than the level of either factor alone (Jensen,1983; Heilbrun et al.,1989). Moreover, it is possible that the effects of fibre must be considered in the context of total diet and of the interactions with various other dietary components.

MECHANISM OF PROTECTION AGAINST COLORECTAL CANCER

by M.J. Hill

When Burkitt (1971) proposed a role for dietary fibre in protection against colorectal cancer, the postulated mechanism was that:

a) Fibre causes stool bulking which dilutes luminal carcinogens;

b) Fibre causes an increased rate of transit through the colon and this decreases the time available for bacterial production of carcinogens and for carcinogen action;

c) Increased colonic carbohydrate provides nutrients for gut bacteria causing a change in the composition and the metabolic activity of the gut flora and a change in the physico-chemical environment in the colon;

d) Metabolism of complex carbohydrates in the colon results in the production of metabolites (particularly butyrate) with antineoplastic properties.

a) Stool bulking. Cereal fibre has potent stool bulking properties with 65-70% increase in stoll mass being produced by 16 g., 39 g. and 54 g. of wheat bran in three separate studies of healthy young English males. These each had a direct diluting effect on two types of stool marker compounds (the acid and the neutral steroids). When a very high dose (100g) of wheat bran was used it produced a tripled stool mass but also caused non-specific washout of all steroids and so no extra dilution over that acheived by half the amount. Oat bran produces a similar stool bulking effect and a similar dilution of neutral compounds. However, oat bran stimulates faecal loss of bile acids and so, despite the stool bulking, the faecal acid steroid concentration actually increases. Similarly fruit fibre (pectin) offers relatively little stool bulking but this is more than offset by a non-specific increase in faecal steroid output leading to an increased faecal concentration of all steroids. Of the fibre sources studied only wheat bran and cellulose have the desired effect of decreasing the stool concentration of both acid and neutral steroids, and of meeting the criteria postulated

b) Rate of transit. While it is well established that cereal fibre increases the rate of colon transit, it is not clear that this results in a decreased extent of bacterial metabolism. The lack of linkage between transit time and extent of bacterial metabolism is well documented (Hill and Fernandez, 1990) and could have many possible explanations.

The suggested relation between transit and mean contact time could be falacious (Hill, 1974). The "faecal stream" is not continuous, slow transit of faecoliths also implies slow transit of the spaces between faecoliths so that the net contact time with faeces is unchanged.

c) Bacterial flora. The relation between diet and the bacterial flora appears self evident so that when during the 1970s and 1980s, repeated studies failed to detect any effect of fibre on the faecal bacterial flora it caused considerable concern. Later studies by Fernandez et al.(1985) and Berghouse et al.(1984) showed, in contrast, a considerable effect of diet on the flora of ileostomy effluent, indicating that the major side of bacterial metabolism of fibre was the proximal colon. In retrospect this should have been predicted. In the proximal colon the fibre concentration would be maximal and the conditions sufficiently fluid to enable maximal metabolism of any substrate, including fibre.

By the time the distal colon and rectum is reached the amount of readily metabolized carbohydrate has been greatly decreased and so its effect on the balance of the flora will have been similarly decreases. This analysis is consistent with the results obtained in canulated pigs by Fadden et al.(1984).

Fermentation of carbohydrate yields volatile fatty acids and this can lead to an acidification of the proximal colon (Bown et al, 1974; Pye,1988). This can be such that putrefactive metabolism may be inhibited and could lead to decreased production of carcinogens and mutagens. This is consistent with the hypothesized mechanism of the protective effect of cereals against colorectal carcinogenesis.

d) Anticarcinogenic metabolites. Much has been made of the possible role of butyrate both as a colonocyte nutrient and as an antineoplastic agent. Carbohydrate is metabolized in the large bowel (particularly in the proximal colon) to volatile fatty acids including butyric acid. Studies by Roediger (1982) showed that butyrate is taken up by colonocytes from the gut lumen in preference to nutrients from the vascular system and this led to a general belief that butyrate was "good" for the colon. The next phase was the demonstration in cultured colonocytes that butyrate can improve cell differentiation and so should have anticarcinogenic properties (eg Gum et al.1987; Kim et al.1980). In contrast, Berry and Paraskevo (1988), while studying markers of progression from adenoma to carcinoma in cultured cell lines, obtained evidence that butyrate can act as a tumour promoter in the colon by favouring growth of malignant clones at the expense of normal tissue. Thus, the evidence from cultured cells is confused; tests in sample systems suggest protection whilst those that attempt relevance to the situation in the large bowel suggest the opposite.

In this situation, studies of patients are relevant. In case-control studies comparing patients with colorectal cancer, or adenomas or controls, if butyrate is protective then colonic butyrate levels should be notably higher in controls than in adenomas; if it is promoting, then the levels should be higher in cancer than in adenoma cases. In fact there was a non-significant tendency for both of these, when faecal butyrate levels were measured. However, these levels are the result of differential rates of production and of absorption between the groups; when production rates were essayed there were no differences among the three patient groups ($p < .40$). Clearly much more work needs to be done if butyrate is to be shown to be responsable for the protective effect of cereals.

So far, it is apparent that knowledge on linkage on fibre and colon carcinogenesis is not satisfactory. Efforts should be made to plan standardized animal model studies and, as far as human studies are concerned, to provide detailed composition in dietary fibre of the most important food items. This will enable the epidemiologist to reanalyse their data and provide a better understanding of the relationship between colorectal cancer and dietary intake of fibre (Faivre,1991).

The relevance of all these data to primary prevention of colorectal cancer will have to be determinated by intervention studies on dietary fibre supplementation. In this context, an intervention study to test the efficacy of dietary supplementation of fibre, or of calcium, on colon adenoma recurrence and on adenoma growth in the large bowel was developed by ECP (European Cancer prevention Organization) (Faivre,1993).

DIETARY FIBRES AND BREAST CANCER

by F. Berrino

The relationship between diet and breast cancer is a very controversial one. On an international level there is a strong positive correlation between breast cancer mortality rates and indicators of average per capita consumption of fat, and also with animal proteins and refined sugars, while the correlation with complex carbohydrates tends to be negative.

Case-control and cohort studies based on the quantification of the individual consumption, on the other hand, have yielded inconsistent results. The issue of fat, in particular, is very controversial. Several studies, however, have suggested that a diet rich in fruit and vegetables is associated with a lower risk of breast cancer; a recent meta-analysis came out with an overall result suggesting that women with a low consumption of fruit and vegetables have a 30% higher risk than those with high consumption (Block,1992).

It is not known, however, what are the relevant substances that confer a protective role to fruit and vegetables. The largest study of diet and cancer so far carried out, the cohort studies of american nurses, is consistent with some protective effect of vegetable foods, but does not suggest any specific protection of vegetable fibres *per se* (Willett et al.,1992). Specific fibres and vegetables might be more protective than others, however. It has been suggested in particular, that soya beans might contribute to the lower risk of breast cancer in eastern countries. A negative association of breast cancer risk with the frequency of consumption of soya foods, has been found in a prospective study in Japan (Hiramama,1985) and in a case-control study in Singapore (Lee,1991). The Japanese study was a very large one with only a few questions on food consumption: the risk of women who had never eaten soybean paste soup, relative to those who ate it daily, was 2.19 (one tailed p=0.024). The Singapore study was based on a complete dietary history and could quantify nutrients from different sources: premenopausal women in the highest tertile of soya protein consumption, with respect to the lower one, had a RR of 0.29 (95% confidence interval 0.15-0.57); no increased risk was detected among postmenopausal women. In both studies, however, it is difficult to exclude that the apparent protection may at least in part depend on the association of soya consumption with a more traditional way of life.

The mechanisms by which dietary fibres may affect the risk of breast cancer have been recently reviewed (Rose,1990; Adlercreutz,1990). Two main hormonal mechanisms have been proposed:

1) Fibres reduce the enterohepatic circulation of steroid hormones by increasing the faecal bulk and changing the intestinal microflora; estrogens are excreted in the intestine through the bile in the form of sulphates and glucuronides; an high fat-low fibre diet would increase the bacterial beta-glucuronidase that makes faecal estrogens reabsorbable.

2) Vegetable fibre rich foods are rich in phytoestrogens (mammalian lignans precursor and isoflavonoids); these are diphenolic substances with feeble estrogenic activity that actually compete with estrogens for their receptors and decrease the overall estrogenic activity in the body; they also increase the level of the Sex Hormone Binding Globulin, thus reducing the availability of sex steroid hormones to their target organs.

REFERENCES

Adlercreutz H (1990). Western diet and western diseases: some hormonal and biochemical associations. Scand J Clin Lab Incest, 50, suppl 201: 3-23

Benito E, Stiggelbout A, Bosch FX, Obrador A, Kaldor J, Mulet M, Munoz N (1991). Nutritional factors in colorectal cancer risk: a case-control study in Majorca. Int J Cancer, 49: 161-167

Berghouse L, Hori S, Hill MJ et al. (1984). Comparison between the bacterial and oligosaccharides content of ileostomy effluent in subjects taking diets rich in refined or unrefined carbohydrates. Gut, 25: 1071-1077

Berry RD, Paraskevo C (1988). Expression of CEA by adenoma and carcinoma derived epithelial cell lines: possible markers of tumour progression and modulation of expression by sodium butyrate. Carcinogenesis, 9: 447-450

Bidoli E, Franceschi S, Talamini R, Barra S, La Vecchia C (1992). Food consumption and cancer of the colon and rectum in north east Italy. Int J Cancer, 50: 223-229

Bingham SA (1990). Mechanisms and experimental and epidemiological evidence relating dietary fibre (non-starch polysaccharides) and starch to protection against large bowel cancer. Proc Nutr Soc, 49(2): 153-171

Block G, Patterson B, Subar A (1992). Fruit, vegetables and cancer prevention: a review of epidemiological evidence. Nutr Cancer, 18: 1-29

Bown RI, Gibson A, Sladen G et al (1974). Effects of lactulose and other laxatives on ileal and colonic pH as measured by radiotelemetry device. Gut, 15: 999-1004

Bristol JB, Emmett PM, Heaton KW, Williamson RCN, (1985).Sugar, fat and the risk of colorectal cancer. Br Med J, 291: 1467-1470

Burkitt DP 1971). Epidemiology of cancer of the colon and rectum. Cancer, 28: 3-13

Cresta M, Lederman S, Garnier A, Lombardo G (1969). Rapport EURATOM.

Dales LG, Friedman GD, Ury HK, Grossman S, Williams SR (1979). A case control study of relationships of diet and other traits to colorectal cancer in American blacks. Am J Epidemiol, 109: 132-144

Eastwood MA, Passmore R (1983). Dietary fibre. The Lancet, 23: 202-205

Englyst HN, Cummings JH (1990). Non-starch polysaccharides (dietary fibre) and resistant starch. In Furda I, Brine CJ, ed. New developments in dietary fibre. New York: Plenum Press., 205-25

Fadden K, Owen R, Hill MJ et al. (1984) steroid degradation along the gastrointestinal tract: the use of the cannulated pig as a model system. Trans Biochem Soc, 12: 1105-1106

Faivre J (1991). Diet and colorectal cancer. In Benito E, Giacosa A, Hill MJ, ed. Public education on diet and cancer. Kluwer Academic Publishers, London, 53-68

Faivre J (1993). The ECP calcium fibre prevention study. Eur J Cancer Prev, 2 suppl.1: 6-7

Fernandez F, Kennedy H, Hill M, Truelove S (1985). The effect of diet on the bacterial flora of ileostomy fluid. Microbiol Aliments Nutrition, 3: 47-52

Ferro-Luzzi A, Sette S (1989). The mediterranean diet: an attempt to define its present and past composition. Eur. J. Clin. Nutr. 43 (Suppl. 2): 13-29.

Fidanza F, Fidanza Alberti A (1971). Rilevamento dei consumi alimentari di alcune famiglie in tre zone regionali d'Italia. Quaderni Nutr. 31: 139-188.

Freudenheim J, Graham S (1989). Toward a dietary prevention of cancer. Epidemiol. Reviews, 11: 229-233

Freudenheim J, Graham S, Marshall JR, Haughey BP, Wilkinson G (1990). A case-control study of diet and rectal cancer in western New York. Am J Epidemiol, 131: 612-624

Graham S, Dayal H, Swanson M, Mittelman A, Wilkinson G (1978). Diet in the epidemiology of cancer of the colon and rectum. JNCI, 61: 709-714

Graham S, Marshall J, Houghey B, Mittelman A, Swanson M, Ziclcany M, Byers T, Wilkinson G, West D (1988). Dietary epidemiology of cancer of the colon in Western New York.Am J Epidemiol, 128:490-503

Gum JR, Kam WK, Byrd JC et al (1987). Effects of sodium butyrate on human colonic adenocarcinoma cells. J Biol Chem, 262: 1092-1097

Heilbrun LK, Nomura A, Hankin JH, Stemmermann GN (1989). Diet and colorectal cancer with special reference to fibre intake. Int J Cancer, 44: 1-6

Hill MJ (1974). Colon cancer: a disease of fibre deplection or dietary excess? Digestion, 11: 293-306

Hill MJ, Morson BC, Bussey HJR (1978). Etiology of adenoma-carcinoma sequence in large bowel. Lancet, 1: 245-247

Hill MJ, Fernandez F (1990). Bacterial metabolism, fiber and colorectal cancer. In Dietary fiber. Kritchevsky D, Bonfield C, Andersons J, ed. Plenum, New York, 417-429

Hill MJ (1991). Dietary fibre and human cancer. In Giacosa A, Hill MJ, ed. The mediterranean diet and cancer prevention. 141-157

Hill MJ (1991). Dietary anti-carcinogens and cancer. In Giacosa A, Hill MJ, ed. The mediterranean diet and cancer prevention. 159-164

Hirayama T (1985). A large scale cohort study on cancer risks by diet with special reference to the risk reducing effects of green-yellow vegetables consumption. Int Symp Princess Takamatsu Cancer Res Fund, 16: 41-53

ISTAT (1981) I consumi delle famiglie, Roma, Istituto di Statistica (Ed.).

Jain M, Cook GM, David FG, Grace MC, Howe CR, Miller AB (1980). A case-control study of diet and colorectal cancer. Int J Cancer, 26: 757-768

Jensen OM (1983). Cancer risk among Danish male Seventh Day Adventists and other temperance society members. JNCI, 70: 1011-1014

Kim YS, Tsao D, Siddigui B et al. (1980). Effects of sodium butyrate and dimethylsulfoxide on biochemical propertiesa of human colon cancer cells. Cancer, 45: 1185-1192

Kune S, Kune GA, Watson LF (1987). Case-control study of dietary etiological factors: The Melbourne colorectal cancer study. Nutr Cancer, 9: 21-42

La Vecchia C, Negri E, Decarli A, D'Avanzo B, Gallotti L, Gentile A, Franceschi S (1988). A case-control study of diet and colorectal cancer in northern Italy. Int J Cancer, 41 (4): 492-498

Lee Hp, Gourley L, Duffy SW, Esteve J, Lee J, Day N (1989). Colorectal cancer and diet in an Asian population: a case-control study among Singapore Chinese. Int J Cancer 43:1007-1016

Lee Hp, Gourley L, Duffy SW, Esteve J, Lee J, Day N (1991). Dietary effects on breast cancer risk in Singapore, Lancet, 337: 1197-1199

Leeds AR (1991). Fibre and resistant starch and cancer. In Benito E, Giacosa A, Hill MJ, ed. Public education on diet and cancer. Kluwer Academic Publishers, London, 85-90

Lyon Jl, Mahoney AW, West DW, Gardner JW, Smith KR, Sorenson AW, Stanish V (1987). Energy intake: its relationship to colon cancer risk. JNCI, 78: 853-861

Macquart-Moulin G, Riboli E, Cornee J, Charnay B, Berthezene P, Day N (1986). Case-control study on colorectal cancer and diet in Marseilles. Int J Cancer, 38: 183-191

Mancini M, Rivellese A, Riccardi G, Postiglione A (1983). Dietary fibers for prevention of CVD. In Atherosclerosis VI. Schettler G, Gotto AM, Middelhof G, Habernicht AS, Jurutla NR, ed. Springer -Verlag, Berlin, 336-340

Martinez I, Torres R, Frias Z, Colon JR, Fernandez N (1981). Factors associated with adenocarcinomas of the large bowel in Puerto Rico. Rev Latinoam Oncol Clin, 13: 13-20

Miller AB, Howe GR, Jain M, Craib KJP, Harrison L, (1983). Food items and food groups as risk factors in a case-control study of diet and colorectal cancer. Int J Cancer, 32: 155-161

Modan B, Barrel V, Lubin F, Modan M, Greenberg RA, Graham S (1975). Low fiber intake as an etiologic factor in cancer of the colon. JNCI, 55: 15-18

Official Methods of Analysis (1990). 15th Ed. AOAC, Arlington, VA, pp. 985-1029.

Potter JD, McMichael AJ (1986). Diet and cancer of the colon and rectum: a case-control study. JNCI, 76: 557-569

Pye G (1988). Gastrointestinal pH and colorectal neoplasia. PhD Thesis University of Nottingham

Rivellese A, Riccardi G, Giacco A, Pacioni D, Genovese S, Mattioli PL, Mancini M (1980). Effects of dietary fibre on glucose control and serum lipoproteins in diabetic patients. Lancet 1: 447-450

Rivellese A, Riccardi G, Giacco A, Postiglione A, Mastranzo P, Mattioli PL (1983). Reduction of risk factors for atherosclerosis in diabetic patients treated with a high fiber diet. Prev Med 12: 128-132

Roediger W (1982). Utilization of nutrients by isolated epithelial cells of the rat colon. Gastroenterology, 83: 424-429

Rose DP (1990). Dietary fiber and breast cancer. Nutr Cancer, 13: 1-8

Saba A, Turrini A, Mistura G, Cialfa E, Vichi M (1990). Indagine nazionale sui consumi alimentari delle famiglie 1980-84: alcuni principali risultati. Rivista della Società Italiana di Scienza dell'Alimentazione, Anno 19, n. 4, pp. 53-65.

Slattery ML, Sorenson AW, Mahoney AW et al. (1988). Diet and colon cancer: assessment of risk by fiber type and food source. JNCI, 80: 1474-1480

Tabelle di Composizione degli Alimenti. Ed. E. Carnovale, C.F. Miuccio. Istituto Nazionale della Nutrizione, Roma, 1989.

Turrini A, Rosolini G, Cialfa E, Carducci S, Di Lena G, Cappelloni M, Lintas C (1992). Dietary fibre intake: trends in Italy, at the "Topics in Dietary Fibre Research" Meeting, Roma-Viterbo, 5-7 May, 1992.

Turrini A, Saba A, Lintas C. (1991). Study of the Italian reference diet for monitoring food constituents and contaminants. Nutrition Research, 11(8): 861-874.

Tuyns AJ, Kaaks R, Haelterman M (1988). Colorectal cancer and the consumption of foods: a case-control study in Belgium. Nutr Cancer, 11: 189-204

Tuyns AJ, Haelterman M, Kaaks R (1987). Colorectal cancer and the intake of nutrients: oligosaccharides are a risk factors, fats are not: a case-control study in Belgium. Nutr Cancer, 10: 181-196

Willett W (1989). The search for the causes of breast and colon cancer. Nature, 338: 389-94

Willett WC, Stampfer MJ, Colditz GA, Rosner BA, Speizer FE (1990). Relation of meat, fat and fiber intake to the risk of colon cancer in a prospective study among women. N Engl J Med, 323: 1664-1672

Willet WC, Hunter DJ, Stampfer MJ, Colditz G, Manson JE, Spiegelman D, Rosner B, Hennekens CH, Speizer FE (1992). Dietary fat and fiber in relation to risk of breast cancer. An 8-year follow-up. JAMA, 268: 2037-2041

Wilpart M (1987). Dietary fat and fibre and experimental colon carcinogenesis: a critical review of published evidence. In Causation and Prevention of colorectal cancer. Faivre J, Hill MJ ed. Elsevier, Amsterdam, p.85-98

Young TB, Wolf DA (1988). Case-control study of proximal and distal colon cancer and diet in Wisconsin. Int J Cancer, 42: 167-175

DIETARY FIBER IN THE PREVENTION OF CARDIOVASCULAR DISEASE

Gabriele Riccardi and Anna V. Ciardullo

Institute of Internal Medicine and Metabolic Diseases
2nd Medical School
University of Naples
Via S. Pansini, 5
80131 Naples, Italy

INTRODUCTION

Atherosclerosis is a major cause of death and disability in the western world. Epidemiological, clinical and pathological studies provide strong evidence for the relationship between diet and atherosclerotic vascular disease. A general consensus has been reached on dietary errors predisponing to premature atherosclerosis in men in order to avoid the insidious progression of atheroma in arterial districts and particularly in the coronaries.

The reduction in dietary cholesterol and saturated fatty acids plays a crucial role in the prevention of atherosclerosis. However, other dietary constituents have also been shown to influence cardiovascular risk factors: dietary fiber, starch, olive oil, omega-3-fatty acids and plant protein; they represent some of the more promising "new" nutritional factors which seem worth consideration in the attempt to prevent atherosclerosis and its complications.

The interest in dietary fiber is based on the observation that several diseases present in western countries (i.e. diabetes, hypertension, coronary heart disease, hyperlipidemia, large bowel diseases and cancer) have infrequent occurence in developing countries. Among the various possibilities to explain this observation the nutritional hypothesis is particularly attractive to explain this difference in disease patterns. In particular, the consumption of unrefined vegetable foods (rich in fiber) is very high in developing countries and low in the western world. In the early sixties Trowell hypothesized for the first time that a low fiber consumption might play a role in the etiology of diabetes, hyperlipidemia and atherosclerosis (1).

In developing the "fiber hypothesis", scientists described potential mechanisms of action. Thus, the theory that a high-fiber diet could prevent certain diseases seemed plausible and stimulated a great deal of research in the following years. So far the final proof of its effectiveness in disease prevention has not yet been obtained. In fact degenerative diseases take many years to develop, and their etiology is multifactorial.

Advances in Nutrition and Cancer, Edited by
V. Zappia *et al.*, Plenum Press, New York, 1993

Table 1. Classification of dietary fiber.

		Fibre	Water solubility
Total dietary fibre	Crude fibre	A) Lignin	-
		B) Non starchy-polysaccharides	
		a) cellulose	-
		b) non cellulose	
		1) Hemicellulose	+ -
		2) Pectic substances	+ + -
		3) Gums	+ +
		4) Mucillages	+ +
		5) Algal polysaccharides	+ +

PHYSIOLOGICAL EFFECTS OF DIETARY FIBER

Fiber represents a heterogenous category and it seems that only some types of fiber are metabolically active. This is probably related to their physical-chemical properties. Some types of fiber are water-soluble and hydrophilic and, therefore, in the presence of water they form a gel. This property influences the process of food digestion since it inhibits the mixing of food with digestive juices thus delaying nutrient absorption. Gums, mucillages, pectines and some hemicelluloses are water-soluble gel-forming and are known to exert useful metabolic effects. They are present in legumes, fruit and vegetables. Conversely, cereals contain mostly unsoluble fiber (lignin, cellulose) and so they are less active on metabolism (Table 1) (2). These considerations can partially explain the different effects of these two types of fiber summerized in Table 2.

The hypolipidemic effects of dietary fiber are probably due to the ability of soluble fiber to bind bile acids thereby increasing their fecal excretion and interrupting the enterohepatic circulation of bile salts; moreover they interfere with intestinal micelle formation and reduce fat absorption (3). Dietary fiber is also extensively fermented by anerobic bacteria in the human colon (4,5). A hypothesis which deserves further studies is that short-chain fatty acids, which represents the fermentative end products of dietary fiber and are absorbed in significant amounts, could contribute to the hypolipidemic and hypoglycemic effect of fiber (6-8). In fact, propionate or acetate lower plasma cholesterol when fed in the diet of experi-

Table 2. Mechanisms of action of dietary fiber.

Water Soluble	Water Unsoluble
* increases fecal bulk	* delays gastric empting
* decreases transit time	* increases transit time
* stimulates mechanical regulation of evacuation	* interfers with nutrients' absorption (carbohydrate, lipids, etc)
* activates bacterial fermentation	* lowers GIP secretion
	* increases somatostatin secretion
	* activates bacterial fermentation
↓ GASTROINTESTINAL EFFECTS	↓ METABOLIC EFFECTS

mental animals and also alter cholesterol synthesis in isolated hepatocytes by competitive inhibition with the 3-OH-3-methylglutaryl-CoA synthetase, which is the first enzyme involved in the synthesis of sterols (9).

Furthermore, dietary fiber has antioxidant properties (10) and this is thought to protect against colon cancer (10) and atherosclerosis. In fact dietary fiber could interfere with in vivo oxidation of lipoproteins. This process makes the lipoproteins cytotoxic to the endothelial cells thus initiating atherogenesis (11). This fascinating hypothesis requires further studies.

EPIDEMIOLOGICAL AND CLINICAL STUDIES

The proven beneficial effects of dietary fiber on two of the most important cardiovascular risk factors (i.e. hyperlipidemia and diabetes) and the suggested useful effects on other risk factors (i.e. hypertension, obesity and thrombogenesis) indicate that high fiber diet can be a useful tool in the prevention of atherosclerosis. Among others, there are two epidemiological studies that show an inverse relationship between fiber intake and coronary heart disease mortality. In particular they show that this relationship is also independent of other dietary variables and cardiovascular risk factors (12,13).

Dietary fiber has an effect on plasma cholesterol in experimental studies. More often the intake of dietary fiber is correlated with plasma cholesterol levels in intercountry comparisons. However many dietary factors affecting plasma cholesterol levels in a similar way tend to cluster together in many diets. Thus, if one compares national diets rich in foods of animal origin and refined cereals with more "vegetarian" diets typical of many developing countries one will observe a clear difference not only in dietary fiber but also in total fat, saturated fat and cholesterol contents. Since all these factors can affect plasma cholesterol concentrations, their combined effects may be important in modifying the rate of progression of atherosclerosis. On the other hand this also makes it difficult to assess quantitatively the effects of the individual dietary factors on atherosclerosis (14).

Population subgroups consuming diets rich in plant foods have lower cardiovascular disease rates than the general population. For example, Seventh-Day Adventists in the Netherlands and Norway have cardiovascular disease rates that are one-third to one-half of those in the general population. Californian Seventh-Day Adventists who eat meat have higher cardiovascular disease rates than do those who are vegetarians. Plasma cholesterol levels among vegetarians are significantly lower than among lacto-ovo-vegetarians and non-vegetarians (14).

Several clinical studies have shown that dietary fiber can be active on some important factors involved in cardiovascular diseases such as carbohydrate and lipoprotein metabolism. Most of them, in fact, have indicated that dietary fiber can improve glucose tolerance and reduce plasmatic concentrations of atherogenic lipoproteins (15,16).

In experimental studies, soluble fiber (pectin, guar gum, etc.) has been shown to reduce blood cholesterol levels, at least when given in large amounts (17,18). Fiber from fresh fruits, vegetables, legumes, oat bran and barley appears to have most potential for reducing cholesterol levels. Except for oats and barley, which are rich in ß-glucans, cereal foods do not appear to have the same potential. Insoluble fiber such as wheat bran has essentially no effect on plasma cholesterol levels (19). In most studies, the fiber used experimentally to lower cholesterol levels was extracted from foods and given as supplement in rather large quantities.

The important role of fiber in the treatment of diabetes, another important risk factor for CHD, was indicated by the pioneer studies of Kiehm et al. (20), who showed a drastic reduction of the insulin dosage and a dramatic improvement of blood glucose control in insulin-treated diabetic patients by prescribing a high-fiber diet. However, from that study, it was unclear to what extent that detected effects could be ascribed to the high-fiber content or to other dietary modifications, including changes in the type and amount of carbohydrate and fat. For this reason we undertook a dietary experiment to evaluate the separate influence of diatery fiber and carbohydrate on blood glucose and lipoprotein metabolism in six IDDM and eight NIDDM patients (15,21). We compared a fat-modified diet in which olive oil (rich in monounsaturated fatty acids) was the major source of fat, with a high-carbohydrate/high-fiber diet in which the carbohydrate enrichment was obtained by increasing the consumption of fiber-rich foods (legumes, fruit, vegetables). In both IDDM and NIDDM patients, the fiber-rich diet produced a significant decrease in postprandial and daily average blood glucose concentrations; fasting glucose levels were also reduced but the difference did not reach statistical significance. Also LDL-cholesterol concentrations were significantly reduced after the high-fiber diet in both types of diabetic patients; an adverse side effect of the high fiber diet was a significant reduction in HDL-cholesterol concentrations in both type I and type II diabetic patients. The results of this study have been confirmed by other groups (22) and concordant data have been obtained in diabetic children, pregnant diabetic women, patients with secondary failure to oral hypoglycemic drugs, and diabetic patients with chronic renal failure (23-26).

In the evaluation of the properties of the high-fiber diet, another aspect that deserves attention is the effect of such a diet on insulin sensitivity which is pathogenetically linked with many risk factors for CHD such as diabetes, hypertension, hyperlipidemia and obesity. In nondiabetic individuals a high-fiber diet induces an improvement of insulin sensitivity (27). In non insulin dependent diabetic patients there are several studies with unclear effects of high-carbohydrate-high-fiber diets on insulin sensitivity (28-29); therefore further investigations are needed to elucidate this aspect of glucose metabolism.

The results of the comparison between a fat-modified and a high-carbohydrate diet go in opposite directions according to whether fiber-depleted starchy foods or unprocessed vegetable foods (fiber rich) are present in the high-carbohydrate diet (Table 3). In the latter case the high-carbohydrate-high-fiber diet has clear metabolic advantages over the fat-modified diet in both IDDM and NIDDM patients. It improves blood glucose control (mainly in the postprandial period), decreases plasma and LDL-cholesterol concentrations, and prevents the elevation of the plasma triglycerides levels, usually observed when dietary carbohydrates are increased. Also, it may improve peripheral insulin sensitivity. Reduction of dietary cholesterol and saturated fat and increased consumption of fiber-rich foods have an additive effect in lowering plasma cholesterol and LDL levels. The overall effect of this combined dietary maneouver is a reduction of plasma cholesterol by as much as 30% in NIDDM patients (15,16,21).

Table 3. Metabolic effects of high-carbohydrate diets in which either starch or fiber-rich foods are mainly represented.

	High Starch Diet	High Fiber Diet
Plasma Glucose	increase	decrease
Plasma Insulin	increase	no effect
Insulin Sensitivity	increase	decrease
Plasma Lipoproteins:		
VLDL	increase	no effect
LDL	no effect	decrease
HDL	no effect or decrease	no effect or decrease

CONCLUSIONS

Diet represents a fundamental tool for cardiovascular disease prevention. Various sets of dietary recommendations have been issued by authoritative international societies to characterize the type of nutrition which is associated with a lower risk for cardiovascular disease (14,18,30).

All these documents agree on the need to prevent or correct obesity reducing daily energy intake and enhancing the level of physical activity, thus improving those risk factors which are often associated with overweight i.e. hyperlipidemia, hypertension, diabetes.

Current dietary recommendations are also concordant on the benefits deriving from a reduction in dietary saturated fat and cholesterol. Saturated fat will be replaced partly by unsaturated fats (mono and polyunsaturated) and partly by complex carbohydrate thus lowering the total fat amount.

Dietary recommendations for cardiovascular disease prevention emphasize the importance of vegetable foods as partial substitutes for foods of animal origin. As previously described, vegetable products are rich in water soluble fiber (especially legumes and fruit) which has an independent cholesterol lowering activity; moreover it improves glucose tolerance and prevents carbohydrate induced hypertriglyceridemia. Another recommendation given in dietary guidelines is to reduce salt and alcohol intake.

Current dietary guidelines are based on a "multifactorial" approach which considers the additive effects of different dietary changes able to prevent atherosclerosis. This means an increased feasibility of the dietary treatment and a higher rate of dietary adherence because our therapeutical objectives can be reached by implementing multiple moderate dietary changes. This more feasible nutritional approach enables the adoption of the same dietary recommendations for high risk individuals and for the general population. In this way the change in patterns of food consumption will be perceived as a major component of a healthy life style rather than a treatment for sick people.

In addition, new dietary guidelines are designed not only to reduce plasma cholesterol but also other important CHD risk factors. The type of diet proposed above has efficacy in lowering blood pressure, improving glucose tolerance, reducing plasma triglyceride and insulin concentrations. Moreover it is considered as an important tool in the prevention of many typical diseases of the western world including stroke, cancer, and bowel diseases.

It can be foreseen that the favorable impact of these nutritional features on CHD and cardiovascular diseases and on total mortality in the population will largely exceed the effects mediated by the reduction of plasma cholesterol levels.

ACKNOWLEDGEMENTS

Supported by Grant n. 91.00226.PF41 from the National Research Council (CNR) - Targeted Project "Prevention and Control of Disease Factors" subproject "Nutrition".

REFERENCES

1. H.C. Trowell. Dietary hypothesis of the etiology of diabetes mellitus. Diabetes 24:762(1975).

2. A. Rivellese and G. Riccardi. Benefits of fibre. Diab. Nutr. Metab. 2(suppl.1):33(1989).

3. J.W. Anderson. Dietary fiber, lipids and atherosclerosis. Am. J. Cardiol. 60:17G(1987).

4. J.H. Cummings, E.W. Pomare, W.J. Branch, C.P.E. Naylor, G.T. MacFarlane. Short chain fatty acids in human large intestine, portal, hepatic and venous blood. Gut 28:1221(1987).

5. J.H. Cummings, N.H. Englyst. Fermentation in the human large intestine and the available substrates. Am. J. Clin. Nutr. 45:1243(1987).

6. J.W. Anderson, S.R. Bridges. Short-chain fatty acid fermentation products of plant fiber affect glucose metabolism of isolated rat hepatocytes. Proc. Soc. Exp. Biol. Med. 177:372(1984).

7. A.O. Akanji, D.B. Peterson, S. Humphreys, T.D.R. Hockaday. Change in plsma acetate levels in diabetic subjects on mixed high fiber diets. Am. J. Gastroenterol. 84:1365(1989).

8. W.L. Chen, J.W. Anderson, D. Jennings. Propionate may mediate the hypocholesterolemic effects of certain soluble plant fibers in cholesterol-fed rats. Proc. Soc. Exp. Biol. Med. 175:215(1984).

9. T. Ide, H. Okamatsu, M. Sugano. Regulation of dietary fats of 3-hydroxy-3-methylglutaryl-CoA reductase in rat liver. J. Nutr. 108:601(1978).

10. Council on Scientific Affairs, American Medical Association. Dietary fiber and health. JAMA 262:542(1989).

11. T.J. Lyons. Oxidized low density lipoproteins: a role in the pathogenesis of atherosclerosis in diabetes. Diabetic Med. 8:411(1991).

12. K. Khaw and H. Connor. Dietary fiber and reduced ischemic heart disease mortality rates in men and women: a 12 year prospective study. Am. J. Epid. 126:1093(1987).

13. L.H. Kushi, R.A. Lew, F.J. Stare et al. Diet and 20 year mortality from coronary heart disease. N. Engl. J. Med. 312:811(1985).

14. WHO Study Group. Diet, nutrition, and the prevention of chronic diseases. Technical Report Series 797:205(1990).

15. A. Rivellese, G. Riccardi, A. Giacco et al. Effect of dietary fiber on glucose control and serum lipoproteins in diabetic patients. Lancet I:447(1980).

16. A. Rivellese, G. Riccardi, A. Giacco et al. Reduction of risk factors for atherosclerosis in diabetic patients with a high fiber diet. Prev. Med. 12:128(1983).

17.Physiological effects and health consequences of dietary fiber. Washington, DC: Life Sciences Research Office, Federation of American Societies for Experimental Biology (1987).

18. Dietary guidelines for healthy Americans: a statement for physicians and health professionals by the Nutrition Commettee, American Hearth Association. Arteriosclerosis 8:218A(1988).

19. G.V. Vahouny. Dietary fiber, lipid metabolism, and atherosclerosis. Fed. Proc. 41:2801(1982).

20. T.G. Kiehm, J.W. Anderson, K. Word. Beneficial effects of a high-carbohydrate, high fiber diet on hyperglycemic diabetic men. Am. J. Clin. Nutr. 29:895(1976).

21. G. Riccardi, A. Rivellese, D. Pacioni et al. Separate influence of dietary carbohydrate and fiber on the metabolic control in diabetes. Diabetologia 26:116(1984).

22. H.C. Simpson, R.W. Simpson, S. Lousley et al. A high carbohydrate leguminous fiber diet improves all aspects of diabetic control. Lancet 1:1(1981).

23. A.L. Kinmouth, R.M. Angus, P.H. Jenkins et al. Wholefoods and increased dietary fibre improve blood glucose control in diabetic children. Arch. Dis. Child. 57:187(1982).

24. D. Ney, D.R. Hollingsworth, L. Cousin. Decreased insulin requirement and improved control of diabetes in pregnant diabetic women given a high-carbohydrate, high-fiber, low-fat diet. Diab. Care 5:529(1982).

25. S.E. Lousley, D.B. Jones, P. Slauyhter. High carbohydrate-high fibre diets in poorly controlled diabetes. Diab. Med. 1:21(1984).

26. M. Parillo, G. Riccardi, D. Pacioni et al. Metabolic consequences of feeding a high-carbohydrate high fibre diet to diabetic patients with chronic kidney failure. Am. J. Clin. Nutr. 48:265(1988).

27. N.K. Fugawa, J.W. Anderson, G. Hageman et al. High carbohydrate, high fiber diets increase peripheral insulin sensitivity in healthy young and old adults. Am. J. Clin. Nutr. 52:524(1990).

28. A. Garg, S.M. Grundy, R.H. Unger. Comparison of effects of high and low carbohydrate diets on plasma lipoproteins and insulin sensitivity in patients with mild NIDDM patients. Diabetes 41:1278(1992).

29. M. Parillo, A.A. Rivellese, A.V. Ciardullo et al. A high-monounsaturated-fat/low-carbohydrate diet improves peripheral insulin sensitivity in non-insulin-dependent diabetic patients. Metabolism 41:1373(1992).

30. Study Group. European Atherosclerosis Society. Strategy for the prevention of coronary heart disease: A policy statement of the European Atherosclerosis Society. Eur. Heart J. 8:77(1987).

CLINICAL RESEARCH AND PERSPECTIVES

DIET AND LARGE BOWEL CANCER

Jean Faivre, Marie-Christine Boutron, Valerie Quipourt

Registre des Tumeurs Digestives
(Equipe associée INSERM-DGS)
Faculté de Médecine
7 Boulevard Jeanne d'Arc
21033 Dijon Cédex, France

Large bowel cancer is one of the most frequent cancers in western countries, thus representing a major health problem. It has been recently estimated in the European Community that the number of new cases was 135,000 each year (30). The prognosis of large bowel cancer remains poor with a 35% 5-year survival rate in population based statistics. Faced with this disquieting situation it seems most likely that primary or secondary prevention will be necessary to control the disease. There is a lot of evidence for attributing most of the differences in incidence between countries to environmental factors, in particular to dietary factors. Migrant studies and studies of religious subgroups with special dietary habits have given considerable support to this hypothesis. The challenge for researchers addressing the relationship between diet and large bowel cancer is to identify causative or protective factors. In this paper possible relationships that relate diet to large bowel carcinogenesis will be analyzed and the usefulness of these data to assist in control and prevention will be underlined.

DIET AND LARGE BOWEL CANCER

Dietary factors that may increase the risk of large bowel cancer

Fat, protein, meat. Internationally the incidence of colon cancer is strongly positively correlated with total fat and total protein intakes and with per capita meat intake. But these studies do not tell us whether the subjects who get colorectal cancer are actually those with the highest fat intake or not.

Things are unfortunately less clear when analyzing case-control studies (Table 1). The association of large bowel cancer with fat intake has been evaluated in seventeen case-control studies: nine found an increased risk for total fat or saturated fat intake (6,14,28,35,39,45,50,58,65), and seven failed to find any such relationship (2,9,22,37,42,62,69). Considering protein intake, seven studies found an increased risk with high protein intake (2, 28, 39, 55, 58, 66) and seven found no effect (6, 19, 22, 35,

Table 1. Main results of case-control studies on suspected risk factors for large bowel cancer.

	Fat	Protein	Meat	Fish	Alcohol	Calories
Haenzel et al (20)			+	-		
Bjelke (4)			+	-	o	
Modan et al (47)						
Philips et al (50)	+		+			
Graham et al (18)			o	+	o	
Dales et al (9)	o		o	o	o	
Martinez et al (45)	+		+			
Haenzel et al (18)			o			
Jain et al (28) Miller et al (46)	+	+	+			
Manousos et al (44)			+	o	o	+
Pickle et al (53)			o	+		
Bristol et al (6)	+	o				+
Potter et al (55)	o	+			+	+
Kune et al (35)	+	o	+	-		o
Macquart-Moulin et al (42), Riboli et al (56)	o	o	o	o	+	o
Tuyns et al (62,63)	o	o	o	o	+	o
Graham et al (19) Freundenheim et al(14)		o		o	o	+
Lyon et al (39)	+	+				+
Slattery et al (58), West et al(64)	+	+				+
La Vecchia et al (36)			+	-	o	
Young et Wolf (69)	o		+			
Heilbrun et al (22)	o	o	o	o		o
Lee et al (37)	o	o	+	o		
Whittmore et al (65)	+	+				+
Benito et al (1,2)	o	+	+	o	o	+
Hu et al (27)			-		+	

+ : increased risks - : decreased risks o : no significant association

37, 42, 62). A high consumption of meat was reported as a risk factor in eleven case-control studies (1,4,20,35,36,37,44,45, 50, 69) over nineteen. Only thirteen studies considered the role of fish in the etiology of large bowel cancer : four studies found a high consumption of fish to be protective (4,20,35,36), whereas one study considered it to be a risk factor (18), and eight studies found no effect (1,9,19,22,37,42,44,63).

In a prospective US study (66), colon cancer was positively associated with beef, pork, lamb, processed meats and liver whereas fish and chicken without skin displayed an inverse relationship. One of the best supports to the high fat diet hypothesis comes from a Canadian study performed in Toronto by Miller and colleagues (28,46). This study, which was very well designed and analyzed from the methodological point of view, reported an increased risk of colon cancer with increasing intake of total fat, total protein and saturated fat. With the multivariate analysis the highest risk was observed for saturated fat with evidence of a dose-response relationship both in males and females. Most studies considering fat as a risk factor were performed in North America or Australia. No association with fat was reported in most West European case-control studies. It is possible that some of the apparent discrepancies among various studies are due to real differences in dietary habits and patterns of risk of various populations rather than to inherent methodological flaws.

Energy. Epidemiological findings on energy intake and cancer have been the object of much controversy. In epidemiologic studies it is difficult to separate the effects of highly correlated factors such as fat, animal protein, meat and total calories. Relatively few case-control studies have considered the association of large bowel cancer with energy intake. Among them, eight found a positive association between total caloric intake and risk of large bowel cancer (2,6,19,39,44,55,58,65), and four found no effect (22,35,42,63). Most studies that found an association with total caloric intake also found an association with fat or protein, while those that did not find an association with calories did not find a association with fat or protein. Any relationship to caloric intake is itself likely to be complex. Exercise has been suggested to have a protective effect against colon cancer, despite the usually increased caloric intake among highly active people. Net energy balance may be an important factor in risk of colon cancer. A role for total calories as an independant risk factor deserves consideration.

Alcohol. An international study in which several environmental variables were correlated with cancer mortality and incidence led to the suggestion that beer consumption might be of etiological importance in rectal cancer. Studies of alcoholics have shown conflicting results. In Denmark no increased risk was found among brewery workers, who on average drink four to five times as much beer as the general population (29). By contrast a similar study of brewery workers in Ireland showed a doubling of the risk of rectal cancer (10). Differences in beer composition that could result from brewing practices or from the components of the beer itself might explain these apparently conflicting results.

Among case-control studies beer was found to be a risk factor for rectal cancer in five studies (15,27,31,35,62), although in one (31) the authors thought the relationship could be due to incomplete control for confounders. In two other studies increased risk with beer consumption was limited to colon cancer. In another study (55) total alcohol intake (but not especially beer) was associated with an increased risk of both colon and rectal cancer. The relationship appeared to be stronger in males than in females. Seven case-control studies have failed to show any association between alcohol and large bowel cancer (1,4,9,18,19,36,44).

In two prospective cohort studies the risk of rectal cancer has been increased with total alcohol use. In one of them, the positive association with consumption of alcohol was limited to beer drinkers whose usual monthly consumption of beer was 15 litres or more (54). The other study suggested that total alcohol use, but no specific beverage type, was associated with increased risk of rectal cancer (33). In a 17-year cohort study in Japan (24) a close association was observed between cancer of the sigmoid colon and alcohol consumption with a relative risk for drinkers vs non drinkers of 4.4 in men and 1.9 in women. The attributable risk in men was estimated as high as 74%. The highest risks were observed for daily beer drinkers but other drinks such as sake and shochu also displayed a strong relationship with the risk of sigmoid cancer.In a cohort of Japanese men in Hawaii (60) rectal cancer was found to be strongly associated with alcohol intake both as total amount and as a percent of total calories and beer was the only alcoholic beverage that displayed a dose-response relationship. Colon cancer was also found to be associated with alcohol but only as a percent of caloric intake. The authors suggested that alcohol might displace cancer inhibitors from the diet.

Dietary factors that may reduce the risk of large bowel cancer

Fiber. Dietary fiber has been suggested as a protective factor for cancer of the large bowel. Burkitt has promoted the idea that the low fiber content of the Western diet would be responsible for the high risk of colon cancer observed in that part of the world (7). This hypothesis is largely based on his observation of low rates of colon cancer in areas of Africa where fiber consumption is high.The results of analytical studies are still rather contradictory. There are seven case-control studies supporting a protective effect of dietary fiber (2,19,35,39,47,58,62), seven studies which found no effect (6,9,22,37,42,46,53), and three studies indicating an increased risk associated with a high fiber intake (45,55,65).

Dietary fiber is not a homogeneous entity and different components may have different physiological effects. Some components may be more efficient than others as protective agents against cancer. The lack of data in food composition tables concerning the different types of dietary fiber may explain that the results of epidemiological studies are rather contradictory. During the last decade many reports have been concerned with the influence of dietary fiber in chemically initiated experimental carcinogenesis mainly in rats. It is apparent from animal studies that attention should be focused on differentiating the types of fibers. They can have an enhancing effect, a protective effect, or no effect on large bowel carcinogenesis depending on the nature, amount, form, and the period of administration (67). Mucilaginous substances and wheat bran appear to be of particular interest (68). Finally, the relevance of all these data to primary prevention of colorectal cancer will have to be determined by intervention studies involving dietary fiber supplementation.

Vegetables. Vegetables have also been a subject of interest in epidemiological studies in relation to large bowel etiology. In most analytical studies, patients with large bowel cancer had a lower consumption of vegetables than cancer-free controls. This was observed in 14 (1,4,18,19,21,27,35,36,42,44,50,63,64,69) out of 21 case-control studies, and in two out of three cohort studies (24,52). Yet the results were somewhat different from one study to another. In the Marseille study the protective effect of vegetables was found for colon cancer, but not for rectal cancer (42). In the Belgium study there was a strong protective effect of vegetables against both colon and rectum cancer and it was stronger for raw vegetables than for cooked vegetables (63). In Western New York significantly reduced risks for colon cancers were observed for high consumptions of tomatoes, peppers, carrots, onions and celery, but not for cruciferous vegetables (19). Yet certain studies showed that the cancer cases had a particularly low intake of cruciferous vegetables.

Only 13 case-control studies considered the role of fruits in the etiology of colorectal cancer. Bjelke found a reduced risk for a high consumption of fruits both in Minnesota and in Norway (4), whereas no significant association was seen in 11 other case-control studies (Table 2).

Vitamins and micronutrients. There has been considerable recent interest in the possible protective effect of vitamins against large bowel cancer. The association of colorectal cancer risk with vitamin A intake was evaluated in eleven case-control studies . No association with risk of colorectal cancer was found in seven (2,6,22,35,36,42,55), and one study found an increased risk in relation to retinol but no association with β carotene (62).

In case-control studies a negative association of vitamin C with colorectal cancer was found in five studies (18,22,35,42,55), but no association was reported in seven other studies (2,6,36,37,46,58,62). There is only limited clinical evidence suggesting the efficacy of ascorbic acid as a chemopreventive agent in patients with polyposis coli (11). No information is available from epidemiological studies concerning Vitamin E.

Recently, a high intake of dietary calcium has also been hypothesized to decrease the risk of colon cancer. In the United States a consistent inverse relationship between calcium intake and colon cancer mortality in the different states was observed both for men and women (30). Epidemiological studies have throroughly addressed the relationship between calcium intake and large bowel cancer. Among ten case-control studies, nine showed no effect of calcium (14,19,35,42,63) and one a protective effect of high calcium consumption (58). Support for the hypothesis was obtained from a 19-year prospective study in the USA (16). In this study the risk of colorectal cancer was inversely correlated with dietary intake of vitamin D and calcium. This result was not confirmed in another cohort study in Hawaii (59). It must be underlined that the method of collecting dietary data differed between the two studies : a 28-day diet record in the first one, a 24-hour recall in the second one.

Enhanced epithelial proliferation in the bowel has been observed in patients at high risk for large bowel cancer. Of interest is the fact that oral calcium supplementation induces a more quiescent equilibrium of epithelial-cell proliferation in the colonic mucosa of subjects at high risk of colon cancer, similar to that observed in subjects at low risk (38,57).

Table 2. Main results of case-control studies on suspected risk factors for large bowel cancer.

	Fibres	Vegetables	Fruits	Vit. A	Vit. C
Haenzel et al (20)		+			
Bjelke (4)		-	-		
Modan et al (47)	-				
Philips et al (50)		-	+		
Graham et al (18)		-			
Dales et al (9)	o				
Martinez et al (45)	+				
Haenzel et al (18)		o			
Jain et al (28), Miller et al (46)	o	o			o
Manousos et al (44)		-	o		
Pickle et al (53)	o	o	o		
Bristol et al (6)	o			o	o
Potter et al (55)	+			o	-
Kune et al (35)	-	-	o	o	-
Macquart-Moulin et al (42),	o	-	o	o	-
Tuyns et al (62,63)	-	-	o	+	o
Graham et al (19) Freundenheim et al(15)	-	-		-	-
Lyon et al (39)	-	+			
Slattery et al (58), West et al(64)	-	-	o	-	o
La Vecchia et al (36)		-	o	o	o
Young et Wolf (69)		-			
Heilbrun et al (22)	o	o	o	o	-
Lee et al (37)	o	-	o	-	o
Whittmore et al (65)	+	+			
Benito et al (1,2)	-	-	o	o	o
Hu et al (27)		-	o		+

+ : increased risks - : decreased risks o : no significant association

Methodological issues

It is apparent that the results of analytical studies on large bowel cancer are not uniform and it is therefore appropriate to discuss the causes of this inconsistency.

-- Studies that fail to show a relation between cancer and a specific nutrient may do so because of the imprecision of the measurements of dietary patterns. Problems are related to the collection of information on a large number of foods or nutrients. There may be patient recall bias influenced by knowledge about the disease, influence of loss of appetite, interaction between various dietary factors. The quality of dietary assessment methods differs from one study to another. In the studies done before 1980 very simple food frequency methods were used to assess diet intake. The validity of collected date vary in relation not only to the method used to collect individual diet but also on the ability of the subject to answer questions and the skill of the interviewer.

-- The lack of variability in diet in most studies may be responsible for the failure to uncover differences in fat and meat intake between cases and control. The association of colon cancer with the consumption of fat and meat was observed in some populations with an assumed wide variation in fat intake such as among the Japanese and Seventh Day Adventists.

-- The imprecision of food tables is another problem. Food composition tables provide an approximation of nutrient intake, particularly for fiber. Composition for different nutrients varies according to the method of analysis. These problems may explain some of the contradictions among studies.

-- Methodological problems are important : the selection of cases or the use of unsuitable controls might also be a factor responsible for the failure to find consistent differences. Confounding factors such as anatomical localization, and metabolic status have not often been taken into consideration. The statistical analysis was often rough. The methodology of the analysis is important. For instance in the Marseille study the final analysis using a logistic model gave results different from a first analysis.

DIET AND LARGE BOWEL ADENOMAS

There is considerable evidence that a high proportion of large bowel cancers develop from a polipoïd precursion lesion, the adenoma.Taking into account epidemiological and histopathological data, in particular those relating the malignant potentiality to size of the adenoma, Hill et al have suggested that large bowel cancer would be a disease wtih a multistep process (23). According to this hypothesis the factors acting at each step of the adenoma sequence i.e., development of a small adenoma, growth towards a large adenoma and carcinoma may be different. A large bowel cancer could possibly be prevented by intervening at any of these stages. This can be achieved only when the factors responsible for the different stages are known. It would be particularly important to identify factors involved in the second step i.e., adenoma growth.

There have been so far only three published studies on the role of diet in the occurence of colorectal adenomas (2,25,43).The first study was carried out on 77 cases and an equal number of controls enrolled in a screening program concerning a population sample of 400 subjects of Norway aged 50-59 (25). The results indicated a higher intake of carbohydrates and fiber for controls and a higher intake of fat for cases. This study was well designed, but has the inconvenient of a rough statistical analysis comparing only mean values, without adjustment for confounders, in particular caloric intake. The second study, performed in Marseille (France), was based on 250 cases and 250 controls (43). The intake of polysaccharides and natural sugar was lower among cases than among controls, the risk of colorectal adenomas decreasing linearly with increasing daily consumption. In contrast sugar added to food and drinks was observed to have the opposite effect i.e., an increasing risk with increasing consumption. The cases also reported a lower consumption of oil, potatoes, K, Mg and vitamin B6. In the third done in Majorca, preliminary results suggest a protective effect of vegetables and an increased risk for consumption of sugar, sweets and pastry (3). Unfortunately none of those studies took into account the multi-stage concept of the adenoma-carcinoma sequence.

The most consistent data on the relationship between colorectal tumors and diet come indirectly from biochemical studies focused on bile acids (61). Metabolic epidemiological and histopathological studies in humans, experimental studies in rodents and *in vitro* studies have provided data relating the faecal bile acid concentration to the risk of large bowel cancer. In some of those, a detailed analysis has been conducted considering the profile of individual faecal bile acids. The ratio of lithocholic to deoxycholic acids has been shown to be higher in cancer cases than in controls. Regarding that ratio significant differences could also be observed between the large adenoma group (>5 mm) and the small adenoma group (49). Such data suggest that the secondary bile acids, lithocholic acid and deoxycholic acid, and mainly their ratio, are important in the step of growth phase of the adenoma i.e., in cancer promotion. They do not seem to play a role in the formation of small adenomas.

Some evidence was published recently on the role of alcohol and tobacco on the risk of large bowel adenomas. In England (19), current smoking was more common in the adenoma group than in the control group and teetotallers were more common in the control group. Both drinking and current smoking led to a RR of 12.7 compared to total abstainers. In Japan (34), the association between adenomatous polyps of the sigmoid and alcohol intake was limited to sake and beer with a dose-response relationship. In the US (33), cumulative smoking and cumulative beer consumption were both associated with adenomas. In the Marseille study total ethanol intake and consumption of wine distillates were not associated with the risk of polyps of the colon or rectum (56). Unfortunately again none of these studies considered the location nor the size of the adenoma. In a study of diet and colorectal tumors set up in Burgundy (5), there was an association between both alcohol and tobacco consumption and the risk of large adenomas whereas tobacco was the main difference between small adenomas and polyp-free controls. A multistage process involving then tobacco for adenoma formation and alcohol for adenoma growth could be proposed. When considering polyp location, tobacco seemed to act at any site of the large bowel whereas alcohol seemed to be mainly involved in the growth of left colon adenomas.

INTERVENTION STUDIES

Because of the inconclusive results of the available studies, there is at the moment a great interest in intervention studies. They are justified when there is enough evidence supporting the hypothesis that a given diet might be of benefit, but on the other hand the evidence must not be so strong as to make it unethical to withold the treatment from the control group. We have seen that there are many dietary factors which lead to plausible mechanistic hypotheses about the role they may play in the development of large bowel cancer. As intervention studies are the most powerful tools in cancer epidemiology to detect small or moderate effects, which can be of importance to decrease the incidence of this illness, it is appealing to investigate some of the available hypotheses within intervention studies.

The end point of such studies cannot be actually invasive cancer itself. They would require a very large number of subjects and an excessively long study period. It is therefore not surprising that most planned or on-going intervention studies are conducted in specific high risk populations, mainly patients with adenomas. Most proposed intervention studies use as the end point the development of new adenomas in populations who initially had a clean bowel following polypectomy. Such studies can only usefully test for factors thought to be implicated in the causation of adenomas. Very few data are available on this step; most available results, particularly those on secondary bile acids, concern adenoma growth. So it is interesting to conduct studies in patients with small adenomas left unremoved in the large intestine. In a large scale longitudinal study conducted in the Oslo area in which small polyps of less than 5 mm in diameter were allowed to remain in the bowel, no cases of severe dysplasia or carcinoma were registered after 2 years (26). Only 40% of the polyps had increased in size and by no more than 3 mm. This indicates that there is no major risk in leaving small adenomas for later follow-up. It provides an ideal model to study the factors determining the growth rates of adenomas i.e., the most important step in the carcinogenesis process.

Preventive measures which should show quick results are mostly directed at promoting factors, because only at this stage alterations in the prevalence of the risk could be effective to stop the carcinogenic process. It is attractive to consider interventions which can be easily implemented by supplementation rather than by changing people's general dietary habits and for which a clear hypothesis deriving from laboratory and epidemiological studies exists. From the bile acids hypothesis a range of possible dietary intervention mechanisms can be formulated.

-- Decrease bile acids concentration entry to the colon by modifications of the diet, particularly by decreasing dietary fat;

-- Dilute the colonic content by stool bulking, using dietary fiber or non-absorbed disaccharides such as lactulose or fructo-oligosaccharides,

-- Decrease the amount of bile acids in solution by precipitating the bile salts for instance with calcium to form calcium soaps;

-- Decrease the metabolism of bile acids in the colon which can be achieved by decreasing the colon pH. Mild acidification of the coecum can be obtained with fiber and lactulose.

Regarding the type of intervention it is of interest to consider an intervention which can be easily implemented and for which a good hypothesis is available. Changing people's dietary habits seems difficult to implement and to evaluate. Giving a dietary supplement is less difficult. To test rapidly some of the available hypotheses a multicentric assay is required. Such a study is under realization within the European Cancer Prevention Organization (ECP). Lactulose and fructo-oligosaccharides were rejected because although they are scientifically interesting they were thought to be unacceptable to patients because of frequent diarrhea or flatulence. Calcium was chosen because it can bind bile acids as soaps and therefore decrease their toxic effect on the colonic mucosa. The fiber hypothesis was also considered; Ispaghula husk, a mucilaginous substance, was chosen because it is known to be an effective stool bulking agent, it decreases caecal pH and it has a potent antitumor activity in animal models during the promoting phase.

The ECP calcium fiber intervention study is complementary to other on-going intervention studies concerning primary prevention of colorectal tumors. It must be underlined that it is one of the few studies together with the Oslo study and the Nottingham study to test the efficacy of a supplement on the growth rate of a adenoma left *in situ* in the large bowel (13). Several studies are currently evaluating the effect of calcium supplementation. The highest dose (2 g/day of calcium) is given in the ECP study. This high dose has been chosen because the aim of the intervention is to maintain the maximum amount of bile acids in aqueous solution. In on-going studies in the USA and UK, calcium is given at a dose of 1 to 1.5 g per day. In the Oslo study calcium is given with selenium, β carotene, vitamin E and vitamin C. Available results from intervention studies concerning vitamins supplements alone are negative (11,40). Only two studies are evaluating the effect of a fiber supplement. In the Australian study, wheat bran has been chosen and in the ECP study a mucilaginous substance.The Australian study is also evaluating the effect of lowering fat (41).

CONCLUSION

Evidence from case control studies or prospective studies provide defensible arguments for dietary implications in the causation of large bowel cancer either as initiators and promoters or as inhibitors of carcinogenesis. There is fairly consistent evidence of the protective effect of vegetables. There is some evidence relating fat intake, fiber intake or calcium intake to cancer of the large bowel. Available data are not sufficient to serve as a basis for strong specific dietary advice and studies on large bowel cancer should be undertaken, particularly in the field of intervention studies.

REFERENCES

1. BENITO E, OBRADOR A,STIGGELBOUT A, BOSCH FX, MULET M, MUNOZ N, KALDOR J (1990) A population-based case-control study of colorectal cancer in Majorca. I. Dietary factors. Int J Cancer 45 : 69-76
2. BENITO E, STIGGELBOUT A, BOSCH FX, OBRADOR A, KALDOR J, MULET M, MUNOZ N (1991) Nutritional factors in colorectal cancer risk : a case-control study in Majorca. Int J Cancer 49: 456-463
3. BENITO E, BOSCH FX, CABEZA E, MORENO V, OBRADOR A, MUNOZ N (1992). Diet, physical activity and colorectal adenomas. A case-control study in Majorca. In Recent progress in colorectal cancer : biology and management of high risk groups. RossiniR, Lynch ht, Winawer S eds. Excepta Medica Amsterdam : 137-141
4. BJELKE E (1974) Epidemiological studies of cancer ot the stomach, colon and rectum ; with special emphasis on the role of diet. Scand J Gastroenterol 9 : Suppl. 31, 1-235
5. BOUTRON MC, FAIVRE J, DOP MC, SENESSE P (1992) L'alcool et le tabac sont-ils aussi à l'origine des cancers colorectaux. Gastroenterol Clin Biol 16 : A185 (Summary)
6. BRISTOL JB, EMMETT PM, HEATON KW, WILLIAMSON RCN (1985) Sugar, fat, and the risk of colorectal cancer. Br Med J 291 : 1467-1470
7. BURKITT D P (1971) Epidemiology of cancer of the colon and rectum. Cancer 28 : 3-13
8. CULLEN JW (1988) The National Cancer Institute's Intervention trials. Cancer 62 : 1851-1864
9. DALES LG, FRIEDMAN GD, URY HK, GROSSMAN S, WILLIAMS SR (1978) A case control study of relationships of diet and other traits to colorectal cancer in American blacks. Am J Epidemiol 109 : 132-144
10. DEAN G, MAC LENNAN R, MC LOUGHLIN H, SHELLY E (1979) The causes of death of blue. collar workers at a Dublin brewery. Br J Cancer 40 : 581-589
11. DE COSSE JJ, MILLER HH, LESSER ML (1989) Effect of wheat fiber and vitamins C and E on rectal polyps in patients with adenomatous polyposis. JNCI 81: 1290-1297
12. FAIVRE J, DOYON F, BOUTRON MC (1991) The ECP calcium fiber polyp prevention study. EuropJ Cancer Prev 1 (suppl 2) : 83-89
13. FAIVRE J, VATN M, ARMITAGE N (1992) A report of the european intervention trials on colorectal cancer prevention. In Recent progress in colorectal cancer: biology and management of high risk group. Rossini FP ed. Elsevier, Amsterdam : 179-186
14. FREUDENHEIM JL, GRAHAM S, MARSHALL JR, HAUGHEY BP, WILKINSON G (1990) A case-control study of diet and rectal cancer in western New York. Am J Epidemiol 131 : 612-624
15. FREUDENHEIM JL, GRAHAM S, MARSHALL JR, HAUGHEY BP, WILKINSON G (1990) Lifetime alcohol intake and risk of rectal cancer in western New York. Nutr Cancer 13 : 101-109
16. GARLAND C, SHEKELLE RB, BARRETT-CONNOR E, CRIQUI MH, ROSSOF AM, PAUL O (1985) Dietary vitamin D and calcium and risk of colorectal cancer : a 19-year prospective study in men. Lancet i : 307-309
17. GIOVANUCCI E, STAMPFER MJ, COLDITZ G, RIMM EB, WILLETT WC (1992) Relationship of diet to rik of colorectal adenoma in men. J Natl Cancer Inst 84 : 91-98.
18. GRAHAM S, DAYAL H, SWANSON M (1978) Diet in the epidemiology of cancer of the colon and rectum. JNCI 61 : 709-714
19. GRAHAM S, MARSHALL J, HAUGHEY B, MITTELMAN A, SWANSON M, ZIELEZNY M, BYERS T, WILKINSON G, WEST D (1988) Dietary epidemiology of cancer of the colon in western New-York. Am J Epidemiol 128 : 490-503
20. HAENSZEL W, BERG JW, KURIHARA M, LOCKE F (1973) Large bowel cancer in Hawaïan Japanese. JNCI 51 : 1765-1799
21. HAENSZEL W, LOCKE FB, SEGI M (1980) A case control study of large bowel cancer in Japan. JNCI 64 : 17-22
22. HEILBRUN LK, NOMURA A, HANKIN JH, STEMMERMANN GN (1989) Diet and colorectal cancer with special reference to fiber intake. Int J Cancer 44 : 1-6
23. HILL MJ, MORSON BC, BUSSEY HJR (1978) Aetiology of adema-carcinoma sequence in the large bowel. Lancet i : 245-247

24. HIRAYAMA T (1989) Association between alcohol consumption and cancer of the sigmoid colon : observations from a japanese cohort study. Lancet ii : 725-727
25. HOFF G, MOEN E, TRYGG K, FROLICH W, SAUAR J, VATN M, GJONES E, LARSEN S (1986) Epidemiology of polyps in the rectum and sigmoid colon. Evaluation of nutritional factors. Scand J Gastroenterol 21 : 199-204
26- HOFF G, FOERSTER A, VATN MH, SAUAR J, LARSEN S (1986) Epidemiology of polyps in the rectum and colon. Recovery and evaluation of unresected polyps 2 years after detection. Scand J Gastroenterol 21 : 853-862
27. HU J, LIU Y, YU Y, ZHAO T, LIU S, WANG Q (1991). Diet and cancer of the colon and rectum : a case control study in China. Int J Epidemiol 20 : 362-367
28. JAIN M, COOK GM, DAVIS FG, GRACE MG, HOWE GR, MILLER AB (1980) A case control study of diet and colorectal cancer. Int J Cancer 26 : 757-768
29. JENSEN OM (1979) Cancer morbidity and causes of death among Danish brewery workers. Int J Cancer 23 : 454-463
30. JENSEN OM, ESTEVE J, MOLLER H, RENARD H (1990) Cancer in the European Community and its member states. Europ J Cancer 26 : 1167-1256.
31. KABAT GC, HOWSON CP, WYNDER EL (1986) Beer consumption and rectal cancer. Int J Epidemiol 15 : 494-501
32. KIKENDALL JW, BOWEN PE, BURGESS MB, MAGNETTI C, WOODWARD J, LANGENBERG P (1989) Cigarettes and alcohol as independent risk factors for -colonic adenomas. Gastroenterology 97 : 660-664
33. KLATSKY AL, ARMSTRONG MA, FRIEDMAN GD, HIATT RA (1988) The relations of alcoholic beverage use to colon and rectal cancer. Am J Epidemiol 128 : 1007-1015
34. KONO S, IKEDA N, YANAI F, SHINCHI K, IMANISHI K (1990) Alcoholic beverages and adenomatous polyps of the sigmoid colon : a study of male self-defence officials in Japan. Int J Epidemiol 19 : 848852
35. KUNE S, KUNE GA, WATSON LF (1987) Case-control study of dietary etiological factors : the Melbourne colorectal cancer study. Nutr Cancer 7 : 21-42
36. LA VECCHIA C, NEGRI E, DECARLI A, D'AVANZO B, GALLOTTI L, GENTILE A, FRANCESHI S (1988) A case-control study of diet and colorectal cancer in northern Italy. Int J Cancer 41 : 492-498
37. LEE HP, GOURLEY L, DUFFY SW, ESTEVE J, LEE J, DAY NE (1989) Colorectal cancer and diet in an asian population. a case-control study among singapore chinese. Int J Cancer 43 : 1007-1016
38. LIPKIN M, NEWMARK H (1985) Effect of added dietary calcium on colonic epithelial-cell proliferation in subjects at high risk of familial colonic cancer. New Engl J Med 313 : 1381-1384
39. LYON JL, MAHONEY AW, WEST DW (1987) Total food intake : a risk factor in colorectal cancer. JNCI 78 : 853-861
40. Mc KEON- EYSSEN, HOLLOWAY C, JAZMAJI V, BRiGHT-SEE E, DION P, BRUCE WR (1988) A randomised trial of vitamin C and E in the prevention of recurence of colorectal polyps. Cancer Res 48 : 4701-4705
41. Mc LENNAN R, BAIN C, MACRAE F, GRATTEN H, BATTISTUTA D, BOCKEY EL, CHAPUIS P, GOULSTON K, LAMBERT J, WAHLQUIST M, WARD M (1991) Design and implementation of the Australian Prevention Project. Front Gastrointest Res 18 : 60-73.
42. MACQUART-MOULIN G, RIBOLI E, CORNEE J, CHARNAY B, BERTHEZENE P, DAY N (1986) Case-control study on colorectal cancer and diet in Marseilles. Int J Cancer 38, 183-191
43. MACQUART-MOULIN G, RIBOLI E, CORNEE J, KAAKS R, BERTHEZENE P (1987) Colorectal polyps and diet : a case-control study in Marseilles. Int J Cancer 40 : 179-188
44. MANOUSOS O, DAY NE, TRICHOPOULOS D, GEROVASSILIS F, TZONOU A, POLYCHRONOPOULOU A (1983) Diet and colorectal cancer : a case control study in Greece. Int J Cancer 32 : 1-5
45. MARTINEZ I, TORRES R, FRIAS Z, COLON JR, FERNANDEZ N (1981) Factors associated with adenocarcinomas of the large bowel in Puerto Rico. Rev Latinoam Oncol Clin 13 : 13-20

46. MILLER AB, HOWE GR, JAIN M, CRAIB KJP, HARRISON L (1983) Food items and food groups as risks factors in a case control study of diet and colo-rectal cancer. Int J Cancer 32 : 155-161
47. MODAN B, BARREL V, LUBIN F, MODAN M, GREENBERG RA, GRAHAM S (1975) Low fiber intake as an etiologic factor in cancer of the colon. JNCI 55 : 15-18
48. MORSON BC, BUSSEY HJR, DAY DW, HILL MJ (1983) Adenomas of large bowel. Cancer Surveys 2 : 451-477.
49. OWEN RW, THOMPSON MH, HILL MJ, WILPART M, MAINGUET P, ROBERFROID M (1987) Importance of the ratio of lithocholic acid to deoxycholic acid in large bowel carcinogenesis. Nutr Cancer 9 : 67-71
50. PHILIPS RL (1975) Role of life style and dietary habits in risk of cancer among Seventh Day Adventists. Cancer Res 35 : 3513-3522
51. PHILIPS RL (1985) Dietary relationship with fatal colorectal cancer among. Seventh-Day Adventist. JNCI 74 : 307-317
52. PHILIPS RL, SNOWDON DA (1983) Association of meat and coffee use with cancers of the large bowel, breast, and prostate among Seventh-Day Adventists: preliminary results. Cancer Res 43 : 2403-2408
53. PICKLE LW, GREEN E, ZIEGLER RG, TOLEDO A, HOOVER R, LYNCH HT, FRAUMENI JF Jr (1984) Colorectal cancer in rural Nebraska. Cancer Res 44 : 363-369
54. POLLACK ES, NOMURA AMY, HEILBRUN LK, STEMMERMANN GN, GREEN SB (1984) Prospective alcohol consumption and cancer. New Engl J Med 310 : 617-621
55. POTTER JD, MAC MICHAEL AJ (1986) Diet and cancer of the colon and rectum. A case control study. JNCI 76 : 557-569
56. RIBOLI E E, CORNEE J, MACQUART-MOULIN G, KAAUS R, CASAGRANDE C, GUYADER M (1991). Cancer and polyps of the colorectum and lifetime consumption of beer and other alcoholic beverages. Am J Epidemiol 133 : 157-166
57. ROZEN P, FIREMAN Z, FINE N, WAX Y, RON E (1989) Oral calcium suppresses increased rectal epithelial proliferation of persons at risk of colorectal cancer. Gut 30 : 650-655
58. SLATTERY ML, SORENSON W, FORD MH (1988) Dietary calcium intake as a mitigating factor in colon cancer. Am J Epidemiol 128 : 504-514
59. STEMMERMANN GN, NOMURA AMY, HEILBRUN LK (1984) Dietary fat and the risk of colorectal cancer. Cancer Res 44 : 4633-4637
60 -STEMMERMANN GN, NOMURA AM, CHYOU PH, YOSHIZAWA C (1990) Prospective study of alcohol intake and large bowel cancer. Dig Dis Sciences 35 : 1414-1420
61 -THOMPSON M, HILL MJ (1987) Etiology and mechanisms of carcinogenesis. Diet , luminal factors and colorectal cancer. In : Faivre J, Hill MJ eds. Causation and prevention of colorectal cancer. Elsevier, Amsterdam : 99-120.
62. TUYNS AJ, HAELTERMAN M, KAAKS R (1987) Colorectal cancer and the intake of nutrients : oligosaccharides are a risk factor, fats are not. A case-control study in Belgium. Nutr Cancer 10 : 185-196
63. TUYNS AJ, KAAKS R, HAELTERMAN M (1988) Colorectal cancer and the consumption of foods : a case-control study in Belgium. Nutr Cancer 11 : 189-204
64. WEST DW, SLATTERY ML, ROBINSON LM, SCHUMAN KL, FORD MH, MAHONEY AW, LYON JL, SORENSEN AW (1989) Dietary intake and colon cancer: sex-and anatomic site-specific associations. Am J Epidemiol 130 : 883-894
65. WHITTEMORE AS, WU-WILLIAMS AH, LEE M, SHU Z, GALLAGHER RP, DENG-AO J, LUN 2, XIANG HUI W, KUN C, JUNG D, TEM CZ, CHENGDE L, YAO XJ, PAFFENBARGER RS, HENDERSON BE (1990) Diet, Physical activity and colorectal cancer among Chinese in North America and China. JNCI 82 : 915-926
66. WILLETT WC, STAMPFER MJ, COLDITZ GA, ROSNER BA, SPEIZER FE (1990) Relation of meat, fat and fiber intake to the risk of colon cancer in a prospective study among women. New Eng J Med 323 : 1664-1672
67. WILPART M (1987) Dietary fat and fiber and experimental colon carcinogenesis : a critical review of

published evidence. In Causation and prevention of colorectal cancer. Faivre J, Hill MJ eds. Elsevier, Amsterdam : 85-98

68. WILPART M, ROBERFROID M (1987) Intestinal carcinogenesis and dietary fibers : the influence of cellulose or fybogel chronically given after exposure to DMH. Nutr Cancer 10 : 39-51
69. YOUNG TB, WOLF DA (1988) Case-control study of proximal and distal colon cancer and diet in Wisconsin. Int J Cancer 43 : 167-175

ON THE NUTRITIONAL ETIOLOGY OF BREAST CANCER

Frits de Waard

Department of Epidemiology
University of Utrecht
The Netherlands

Despite immense research efforts, the causes of breast cancer are still incompletely understood. Still, it would be too pessimistic to say that there is no insight at all. Epidemiologists have pinpointed a number of risk factors and experimental medicine has provided not only the understanding that carcinogenesis consists of a series of steps but also that endogenous (hormones) as well as exogenous (e.g. nutrition) factors are involved in the aetiology of breast cancer.

What remains to be elucidated, is the way such factors interact in humans. In order to design preventive strategies it is imperative to have knowledge on the mechanisms involved. A reasonable hypothesis is that the hormonal profile of individuals or even groups is influenced by their life-style, viz. nutritional habits and reproductive experience.

In the past the author advanced the hypothesis that obesity could be the nutrition-related factor which influenced the hormonal profile (1). This idea arose from studying the epidemiology of endometrial cancer in which the triad: obesity-hypertension-diabetes and the unopposed action of oestrogens had been identified as (potential) risk factors. Cytohormonal studies in well-women at postmenopausal age showed indeed that oestrogen levels correlated positively with the degree of overweight (2). The mechanism involved at the molecular level was later shown to be the conversion of the steroid androstenedione into oestrone in adipose tissue (3).

Application of this notion led to the distinction between premenopausal and postmenopausal breast cancer. In a number of case-control studies (4.5) it was found that an association between obesity and postmenopausal (but not premenopausal) breast cancer existed. A complication was that body height appeared to be a risk factor as well (6) and correcting weight for height tended to decrease (or even abolish) the statistical effect of overweight.

A milestone in the understanding of the role of obesity was the discovery by Donegan et al. (7) that it was a prognostic factor, recurrence-free time being adversely affected by its presence. This finding was confirmed by a dozen others (for references see 8) and the insight was gained that obesity is probably a factor which acts late in the natural history of breast cancer, viz. through its effect on extra-ovarian oestrogen production it is a stimulus for oestrogen-dependent breast cancer at a time when ovarian steroid production has come to an

Advances in Nutrition and Cancer, Edited by
V. Zappia *et al.*, Plenum Press, New York, 1993

end. Recently we have concluded a feasibility study on weight reduction in breast cancer patients assuming that such intervention should potentially improve prognosis of patients after menopause (8).

In the United States a strong case is being made for an effect of dietary fat on mammary carcinogenesis (9). The evidence is partly based on international correlations between fat intake and breast cancer incidence in humans, and partly on animal experiments. Concerning the former it should be made clear that several other parameters of Western nutrition and life-style correlate equally well. Concerning the latter the high energy value of fats causes problems of collinearity. Moreover, the work by Pariza (10) points to the fact that the energy provided by fats might be larger than if one does the calculations on the basis of the classical Atwater values (9,4 and 4 kcal per gram of fat, protein and carbohydrate respectively). Thus one could challenge the way in which isocaloric diets have been constituted, the implication being that effects thus far ascribed to fat may be due to its caloric value.

When the fat hypothesis was tested in women by means of case-control studies only meager odds ratios were found. Cohort studies provided no evidence whatsoever for an effect of dietary fat on breast cancer risk. Thus, a critical overview by Goodwin and Boyd (11) expresses doubt as to the role of dietary fat eaten during adulthood. Nevertheless, Prentice (12) lists reasons for believing that the hypothesis is still worth testing by means of an intervention study. The cost, however, of such effort will be enormous.

Over the past eight years I have become interested in the possibility that nutritional factors may have an effect on mammary carcinogenesis much earlier in a woman's life. There is a variety of established facts which suggest such a role. They are the following:

1. **Descriptive epidemiology** tells us that:

a. Japanese migrants to the United States did not get the high breast cancer incidence which prevailed in their new environment. However, their children and grandchildren experienced a strong rise in incidence (13). The old generation either kept its previous life-style or was in a way immune to the influence of American life.

b. In Iceland the transition from a low incidence of breast cancer to a high one during the 20th century took place cohort-wise, viz. each generation experienced higher risk than the previous one (14).

c. Time trends of breast cancer mortality in the United States also suggest that analysis by birth cohort provides clearer trends than analysis by age only (15). Again, this points to the importance of events which take place early in life.

2. **Analytic epidemiology** has revealed the existence of two risk factors occurring at a young age which undoubtedly have to do with nutrition: an early age at menarche and tall height. Rich nutrition as is observed in countries of "Western" life-style apparently has an effect on hormonal profiles during puberty and adolescence.

It should be pointed out that an early menarche not only means earlier exposure to effective oestrogen levels but also to higher levels during postmenarcheal years (16). Therefore it can be inferred that proliferation of oestrogen-dependent tissues (like mammary epithelia) is more intense in populations with Western nutritional patterns.

The significance of this becomes even more clear when one combines the effect of nutrition with that of reproductive trends in the Western world. The risk of having a first full-term pregnancy relatively late in life has been found in many studies (17). Apparently some biologic indicator of risk develops during adolescence and early adulthood and is being stopped (at least temporarily) by the hormones of pregnancy. Experimental work by Russo et al. suggests that it is the differentiation of mammary epithelia brought about by the hormones of pregnancy which renders the breast less sensitive to carcinogenesis (18).

The concept of preneoplastic lesions has its morphological expression in the form of atypical hyperplasia of small ducts or tubulo-ductulo-lobular units (19). In humans evidence is mounting that such lesions begin to develop already at a young age, viz. in the period between menarche and the birth of the first child.

The scientific reasoning is based on two findings:

1. low parity and delayed childbirth are associated with the risky mammographic patterns DY and P2 (20);

2. these patterns are associated with histologic changes, viz hyperplasia without and with atypia the latter being a precancerous condition (21).

Therefore, the life-style as seen in developing countries characterised by restricted food intake, late menarche, early first childbirth and high parity is to protect against the occurrence of preneoplastic changes in the breast (22).

If this hypothesis is correct it has important research and public health implications. If breast cancer prevention is to be addressed at the roots of carcinogenesis, methods of a non-invasive nature should be developed to study the adolescent human breast. This will enable us to focus on the critical period in a woman's life in which the interplay between nutritional and reproductive factors determines the occurrence of lesions that will ultimately lead to breast cancer.

If future research were to confirm that preneoplastic lesions in the breast begin to build up already during adolescence and early adulthood, preventive approaches could focus on changes in life-style through health education.

However, if advice concerning the potential preventive value of low-energy diets during adolescence and early motherhood would be unacceptable socially, one might consider other means. The author is of the opinion that research on a special form of chemoprevention aiming at early differentiation of mammary epithelia should be seriously considered.

REFERENCES

1. F. de Waard, The epidemiology of breast cancer: review and prospects, *Int. J. Ca* 4: 577-586 (1969).
2. F. de Waard, E.A. Baanders-van Halewijn, Cross-sectional data on oestrogenic smears in postmenopausal women, *Acta Cytol. (Philad.)* 13: 675-678 (1969).
3. P.C. Mac Donald, C.D. Edman, D.L. Hemsell, J.L. Porter, P.K. Siiteri, Effect of obesity on conversion of plasma androstenedione to estrone in postmenopausal women with and without endometrial cancer, *Amer. J. Obst. Gynecol.* 130: 448-453 (1978).
4. A.P. Mirra, P. Cole, B. Mac Mahon, Breast cancer in an area of high parity. Sao Paulo, Brazil. *Cancer Res.* 31: 79-83 (1971).
5. F. de Waard, Nutritional etiology of breast cancer: where are we now and where are we going? *Nutrition and Cancer* 4: 85-89 (1982).
6. F. de Waard, E.A. Baanders-van Halewijn, A prospective study in general practice on breast cancer risk in postmenopausal women. *Int. J. Ca* 14: 153-160 (1974).
7. W. Donegan, A.J. Hartz, A.A. Rimm, The Association of body weight with recurrent cancer of the breast, *Cancer* 41: 1590-1594 (1978).
8. F. de Waard, R. Ramlau, Y. Mulders, T. de Vries, S. van Waveren, A feasibility study of weight reduction in obese postmenopausal breast cancer patients, *Eur J. Ca Prev.* (to be published, 1993).
9. News, Low-fat diet trial set to take off, *J. Nat. Ca Inst* 82: 1736 (1990).
10. M.W. Pariza, Dietary fat, caloric restriction, ad libitum feeding and breast cancer risk. *Nutr. Rev.* 45: 1-7 (1987).
11. P.J. Goodwin, N.F. Boyd, Critical appraisal of the evidence that dietary fat intake is related to breast cancer risk in humans, *J. Nat Ca Inst.* 79: 473-485 (1987).

12. R.L. Prentice, M. Pepe, S.G. Self, Dietary fat and cancer: a quantitative assessment of the epidemiological literature and a discussion of methodological issues. *Cancer Res.* 49: 3147-3156 (1989).
13. J. Dunn, Breast Cancer among American Japanese in the San Francisco Bay area, *Nat Ca Inst. Monogr.* 47: 157-160 (1977).
14. O. Bjarnason, N.E. Day, G. Snaedal, H. Tulinius, The effect of year of birth on the breast cancer age curve in Iceland, *Int. J. Ca* 13: 689-696 (1974).
15. R.E. Tarone, K.C. Chu, Implications of birth cohort patterns in interpreting trends in breast cancer rates, *J. Nat Ca Inst.* 84: 1402-1410 (1992).
16. D. Apter, M. Reinila, R. Vihko, Some endocrine characteristics of early menarche, a risk factor for breast cancer, are preserved into adulthood. *Int J. Ca* 44: 783-787 (1989).
17. B. Mac Mahon, P. Cole, T.M. Lin et al., Age at first birth and breast cancer risk, *Bull W.H.O.* 43: 209-221 (1970).
18. J. Russo, L.K. Tay, I.H. Russo, Differentiation of the mammary gland and susceptibility to carcinogenesis, *Breast Ca Res. Treat* 2: 5-73 (1982).
19. S.R. Wellings, H.M. Jensen, R.S. Marcum, An atlas of subgross pathology of the human breast with special reference to possible precancerous lesions. *J. Nat Ca Inst.* 55: 231-273 (1975).
20. F. de Waard, J.J. Rombach, H.J.A. Collette, B.J. Slotboom, Breast cancer risk associated with reproductive factors and breast parenchymal patterns. *J. Nat Ca Inst.* 72: 1277-1282 (1984).
21. N.F. Boyd, H.M. Jensen, G. Cooke, H. Lee Han, Relationship between mammographic and histological risk factors for breast cancer, *J. Nat Ca Inst.* 84: 1170-1179 (1992).
22. F. de Waard, Preventive intervention in breast cancer, but when? *Eur J Ca Prev.* 1: 395-399 (1992).

DIET AND GASTRIC CANCER

Peter I Reed

Lady Sobell Gastrointestinal Unit
Wexham Park Hospital
Slough, Berkshire
SL2 4HL, United Kingdom

It is now widely recognised that diet is important in the aetiology of the intestinal type of gastric cancer (GC) especially as there has been a progressive reduction of GC during the past 40 years even though numerically it is still the second most common cancer world-wide. In developing countries GC ranks second after cervical cancer, and in developed countries fourth after lung, colorectal and breast cancer. There are large differences in incidence among populations with a two to three fold excess in males and incidence increases with age in both sexes. Inverse socio-economic, as well as North-South geographical gradients have been observed in most populations in the Northern hemisphere[1]. An increased risk has been linked to certain occupations including coal mining, fishing and agriculture, and since occupations are clearly linked to socio-economic status some of the observed excess risk might be attributable to dietary habits.

The gradual decline in GC in many populations may be a reflection of the improving economic conditions. Studies in migrants by Haenszel and colleagues in the USA[2] have also helped enormously in our understanding of the dynamics of GC and its precursor lesions. They showed that migrants from a high risk country, Japan, moving to low risk areas, Hawaii and California, retained the high risk for the disease unless they migrated in childhood or early adult life[3]. It is most probable that these differences are related to a change in diet. These and other published studies have provided convincing evidence of the critical importance of exposure in early life in determining the risk of developing, as a first stage, the precancerous lesions and later GC itself.

Extensive epidemiological studies in Japan, which until recently had the highest GC incidence in the world, have highlighted the importance of diet and nutrition in GC incidence[4]. The Japanese National Nutritional Survey demonstrated that striking changes had taken place in the Japanese diet following World War 2. For instance, the consumption of milk and milk products between 1949 and 1978 had increased 28-fold and the ratios of increases in other food items included meat 12.8, eggs 13.0, oil 10.2 and fruit 6.6. The decreasing ratio in the age adjusted death rates for GC in 12 Japanese districts correlated closely with fat and vitamin A intake. In another large prospective study of 265,118 adults aged 40 or over, during the period 1966-1978, Hirayama and colleagues[5] showed a

Advances in Nutrition and Cancer, Edited by
V. Zappia *et al.*, Plenum Press, New York, 1993

significant inverse trend between GC incidence and intake of green-yellow vegetables in both males and females and smokers and non-smokers.

Lauren's description of two major histological types of GC[6] the diffuse and intestinal, has led to a clearer understanding of the epidemiological changes described as it is the intestinal type, found predominantly in the antrum and body of the stomach, which has been decreasing steadily, whereas the incidence of the diffuse type, evenly divided between the sexes, has remained relatively constant in all populations over the years.

Correa *et al* in 1975[7] proposed a model for intestinal type gastric carcinogenesis suggesting a multistage process probably initiated early in life in which the first stage is multifocal chronic atrophic gastritis progressing to intestinal metaplasia (IM) then to epithelial dysplasia and ultimately GC. In the light of increasing knowledge this model has been updated most recently in 1991[8].

In it was proposed that the reduction of nitrate (NO_3) to highly reactive nitrite (NO_2) formed through the action of nitrate-reducing bacteria in the saliva and an hypoacidic stomach could result in the subsequent formation of N-nitroso compounds (NOC) which could act as promoters during the later stage of carcinogenesis and that their effects could also be modified by a number of environmental factors (see Figure 1).

However, the endogenous formation of NOC is not confined to bacterial action alone, either through chemical nitrosation with bacterially formed NO_2 or by bacteria themselves, since N-nitrosation may also occur under normal gastric (acidic) conditions. That *in vivo* N-nitrosation occurs in man has been convincingly confirmed by Ohshima and Bartsch[9] who showed that N-nitrosoproline formed in the stomach following the ingestion of proline and NO_3 is excreted in the urine in inverse ratio to gastric pH and also that this reaction can be inhibited by antioxidants, such as vitamin C. N-Nitroso compounds (NOC), several hundred of which have now been identified, have been shown to be carcinogenic in all 40 animal species tested and it is unlikely that man would be an exception[10 & 11].

The association between NO_3 exposure and GC risk has been extensively studied and the data reviewed by Hill[12] and Forman[13]. Numerous reports have shown that populations at high GC risk had a high NO_3 content in drinking water or in the diet or a high exposure to NO_3 fertilisers. In 1983 Hartman reported a significant correlation between daily NO_3 intake and GC mortality in 12 countries[14].

Some more recently reported studies could not confirm these positive findings[15-17]. The evidence for and against has been reviewed by Preussmann and Tricker[18] . It should be

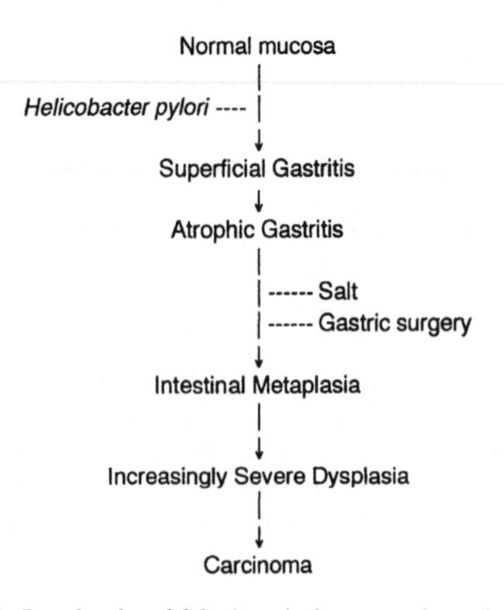

Figure 1. Postulated model for intestinal type gastric carcinogenesis

noted that while NO_3 *per se* does not constitute a cancer risk, it acts as a precursor to NO_2, formed in the body by bacterial reduction, which reacts directly or indirectly to produce NOC. Thus the nitrate burden should be regarded as an "indirect risk" factor and NOC formation as a "direct risk" factor in human carcinogenesis. The importance of dietary factors, has been highlighted by several Far Eastern studies. In 1982 Xu showed in a Chinese study that the NO_3 and NO_2 burden from vegetables and drinking water was higher in an high risk area for GC[19]. Higher salivary and gastric juice NO_3 and NO_2 levels were also demonstrated in fasting patients with chronic gastritis.

Recently, in a second study also in China, a different Xu and colleagues[20] analysed a large number of samples of drinking water in a high risk GC area of North Eastern China for NO_3 and NO_2 and examined the relationship between gastric mucosal lesions and quality of the local drinking water (Table 1) and NO_3 intake via water (Table 2). The staple vegetables consumed by the local population are the high NO_3 containing Chinese cabbage and turnips. Further, a type of salt-turnip consumed daily a few years ago has been shown to contain several volatile N-nitrosamines.

In a third study in Japan[4] the NO_3 burden, determined by urinary excretion, was higher in the low GC risk area and correlated well with vegetable consumption. This would suggest that the diet and dietary sources of NO_3 may be more significant than the total NO_3 burden. The presence of N-nitrosation inhibitors, such as vitamin C and phenolic compounds, which also occur in vegetables and reduce endogenous N-nitrosation, and thereby the GC risk, in low risk populations which show high salivary NO_3 levels, may also be relevant. Large variations in the intake of nitrate both within and between populations were confirmed in a recent ECP-Intersalt collaborative study of 24 hour urinary NO_3 excretion in 5,700 subjects in 48 populations from 28 countries[21]. This was due, in part to the type of diet, notably its vegetable content. NO_3 excretion was also found to he higher in low income populations, e.g. parts of Mexico and Eastern Europe, in the age group 20 - 29 years. On the other hand,

Table 1. Relationship between water type and gastric histology in Moping county, China.

Subject Group (n)	Sex M:F	Age Mean ± SD	Salt %	Mild-salty %	Sweet %
Normal & superficial gastritis (30)	16:14	53.3 ± 10.3	3.3	23.7	73.3
Intestinal metaplasia (32)	16:16	53.3 ± 10.8	18.8	28.1	53.1
Gastric cancer & dysplasia (30)	16:16	53.8 ± 10.3	56.7	20.0	23.3

Table 2. Nitrate and nitrite concentrations in drinking water and estimated nitrate intake of subjects from a high-risk area for GC in Moping county, China.

Subject Group	Nitrate mg/l mean ± SD	Nitrite mg/l mean ± SD	Estimated nitrate intake mg/day
Normal & superficial gastritis	60.1 ± 39.3	0.012 ± 0.020	90.2 ± 59.0
Intestinal metaplasia	72.1 ± 48.6	0.009 ± 0.020	108.2 ± 72.9
Gastric cancer & dysplasia	138.2 ± 101.2	0.018 ±0.032	207.3 ± 151.8

in a study in the UK Forman found an inverse relation between NO_3 intake and urinary levels and GC risk[15]. Therefore, a review of all available data would indicate that in developed countries there is only weak evidence for dietary NO_3 involvement in GC development, whereas in developing countries, such as China and Colombia, the situation is more complex in view of the high incidence of gastric hypoacidity and dietary deficiencies; a high NO_3 burden may thus be but one of several factors responsible for the higher GC incidence.

The role of amine precursors to NOC relevant to human GC induction has also been studied in various countries, again with conflicting results, and the lack of knowledge about specific amine precursors to potentially relevant gastric carcinogens is a major gap in our understanding of human gastric cancer formation.

Positive epidemiological studies have emphasised the nutritional status and specific food items. For instance, smoked fish has been positively associated with GC risk in Iceland[22], Japan[23] and Norway[24], but not in Colombia[25]. Also following nitrosation, a common fish constituent of the Japanese diet, was shown to induce adenocarcinoma in the glandular stomach of experimental rats[26]. Sun-dried herrings and sardines were shown to have strong mutagenic activity after NO_2 treatment[27].

Epidemiological data from Colombia[25] have shown an high intake of fava beans in high GC risk populations. A very potent mutagen, whose precursor was identified as 4-chloro-6-methoxyindole, is produced following nitrosation of fava beans[28]. Other observations suggest that dietary substituted indoles, widely abundant in green plants of the brassica family, can also react with NO_2 in the human stomach resulting in DNA damage[29].

Dry heating (pyrolysis) of food products rich in amino acids and proteins produces highly mutagenic aromatic amine compounds[30]. Certain Japanese foods, including grilled beef, chicken and onions and boiled dried fish have been shown to contain between 1-650 $\mu g/kg$ of pyrolysis products, and the most important representatives of this group of compounds have been shown to be carcinogenic in several animal species[31]. Thus, in the presence of nitrosatable amine precursors the potential formation of NOC and/or alkylating species, which may be relevant for GC induction, remains a plausible working hypothesis. However, the occurrence and nature of nitrosatable amines in the human stomach remains one of the major missing links in our knowledge about human endogenous NOC formation.

Another factor related to diet that is known to have made a large impact on GC is the introduction of refrigeration. There has been a consistent relationship between the start of the decrease in the incidence of GC and the use of refrigeration using ice-boxes, refrigerators and deep-freezes in Japan[32], Spain[33] and USA. Refrigeration makes the use of salt as a food preservative redundant thereby reducing the GC risk and also decreasing the likelihood of growth of moulds in foods[34].

Furthermore, salt itself has been suggested as being associated with GC in its role as a gastric irritant[35]. Several case-control studies, although fairly subjective, have shown remarkable consistency indicating that heavy use of salt would be compatible with a 50% increase in GC risk.

That preserved food intake causes an increased GC risk has also been established in numerous case-control studies world wide[13].

The role of micronutrients and vitamins as inhibitors of N-nitrosation in the aetiology of GC has also been widely studied and reference to this has already been made. There is well documented epidemiological evidence that vegetable consumption does have a protective effect on GC risk[13, 36-41]. This effect has been demonstrated either in developing countries or high risk areas. However, one case control study in the UK[42] failed to show any effect of vegetable consumption on GC risk. A similar protective effect on GC risk has also been shown for fruit[13].

A prospective study carried out in 2,974 healthy men in Basle, Switzerland in whom blood vitamin and micronutrient levels were measured regularly over several years[43] revealed that plasma vitamin A levels were significantly lower in men who subsequently died of stomach cancer. The authors also suggested a new parameter called the normative cumulative index which sums the antioxidant vitamins A, C and ß-carotene and when age-standardised was shown to be significantly depressed for all major cancer types. These findings have been confirmed by an updated analysis at a 12 year follow up which found that stomach and lung cancer were associated with low mean plasma carotene levels ($P<0.001$) and that subjects with subsequent stomach cancer also had lower mean vitamin C and lipid adjusted vitamin A levels than survivors[44].

It was in the early 1980's that the ECP-EURONUT- IM project was conceived as a collaborative study within the framework of the joint IM project of the Diet and Cancer group (subsequently renamed the Upper Digestive Tract Cancer Group) of the ECP (European Cancer Prevention Organisation) , and EURONUT, which is the concerted action programme of nutrition within the European Community[45]. The aim of the study was to establish whether there was a significant difference between intestinal metaplasia cases and age and sex matched controls in the intake of foods, nutrients and toxicants with special emphasis on those which have been postulated as being associated with chronic atrophic gastritis, intestinal metaplasia or gastric cancer. Three groups of patients (aged 20-55) were studied; -those with histologically proven intestinal metaplasia (IM) were age (± 3 years) and sex matched with endoscopic controls (EC) who had histologically normal gastric mucosa and non-endoscopic controls (NC) recruited from surgical out patient departments and who had no history of upper gastrointestinal disease.

Preliminary results[46] have shown that patients in several countries including UK, Poland and Greece with intestinal metaplasia (IM) who have a diet low in intakes of fruit and vegetables and that IM patients in the UK have lower vitamin C plasma levels than either group of controls and that the levels are significantly different between centres[47] (Figure 2).

In this context one must not overlook the role played by *Helicobacter pylori* infection of the stomach, whose prevalence is related to socio-economic factors and age[48]. Childhood overcrowding and large family size are undoubtedly a major factor in the developing countries where 85-90% of the population may be infected. Many recent studies have focused on the relationship between *H.pylori* and IM[49]. Both factors are associated with GC. There is general agreement that IM occurs when gastric atrophy is present and that this condition most commonly results from long term *H.pylori* associated gastritis. *H.pylori* may also impair gastric anti-oxidant defences as low gastric juice AA levels have been demonstrated in *H-pylori* infected subjects[50] and in the presence of hypochlorhydria[51]. Furthermore, eradication of *H.pylori* infection leads to a return to normal gastric juice AA levels[52].

We have studied the effect of vitamin C administration in patients at high risk of developing GC[53] . Those who had undergone gastric surgery for benign peptic ulcer disease had significantly reduced plasma vitamin C levels. Four weeks treatment with 4 g vitamin C daily resulted in a significant reduction in median NOC and NO_2 concentrations in fasting gastric juice.

Krytopoulos[54] showed that supplementation of a normal diet of subjects with 400 mg daily each of vitamin C and alpha-tocopherol resulted in a significant reduction in the mutagenic compounds excreted with faeces suggesting that antioxidants in the diet may have a role in lowering the body's exposure to endogenously formed mutagens.

In a third study Correa (personal communication) not unexpectedly was unable to show any effect on gastric histology in eighty patients with atrophic gastritis after two months treatment with either vitamins A, C or E compared to placebo but was able to demonstrate good compliance.

Although the wealth of available epidemiological case-control and preliminary intervention study data have clearly established the dietary impact on GC risk, there is no doubt that the development of GC is multifactorial. It is now possible to identify at least

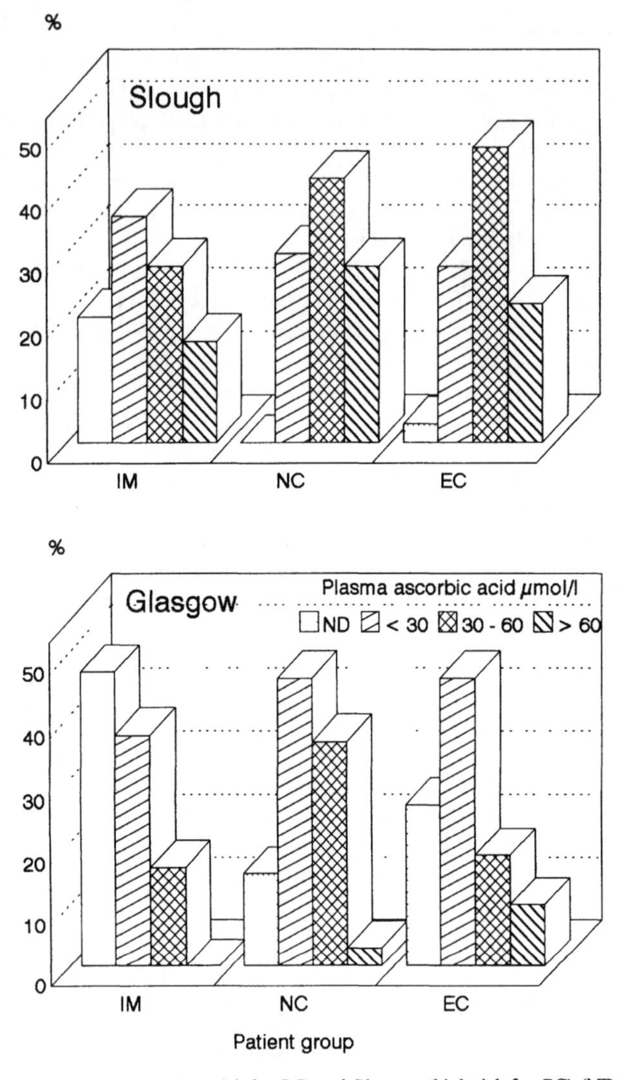

Figure 2. Vitamin levels in Slough (low risk for GC) and Glasgow (high risk for GC) (ND = not detectable)

three factors which probably play a significant role in increasing the GC risk. Two of these are dietary, excessive salt intake and a low intake of fresh fruits and vegetables, and the third is *Helicobacter pylori* infection.

Thus the diet of populations at high risk of GC is characterised by a low intake of animal fats and proteins, high intakes of starches and carbohydrates (mainly from grains and starchy roots), high salt intakes, high dietary nitrate intakes (from water and foodstuffs) with low intakes of fresh fruit and raw vegetables and salads. We await the outcome of future formal studies which, hopefully, will confirm that minimising these risk factors should lead to a further reduction in the incidence of intestinal type GC world wide.

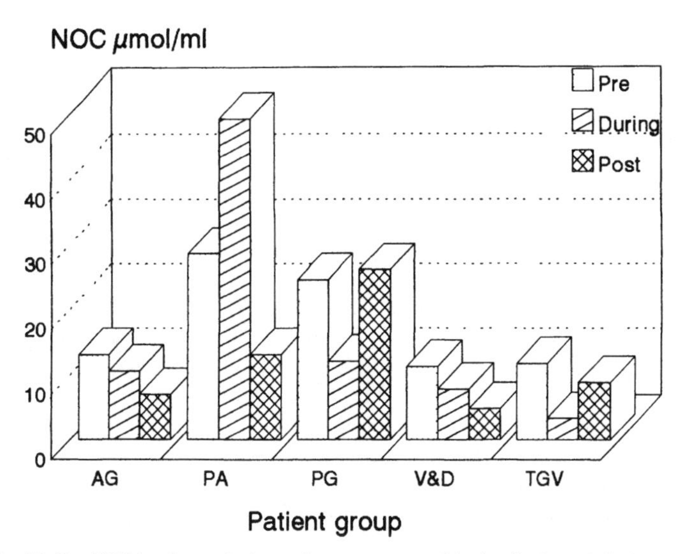

Figure 3. Median NOC levels pre, during and post treatment with vitamin C 4 g daily for 4 weeks. (AG = atrophic gastritis [n=7], PA = pernicious anaemia [n=13], PG = partial gastrectomy [n=10], V&D = vagotomy & drainage [n=26], TGV = truncal vagotomy [n=25])

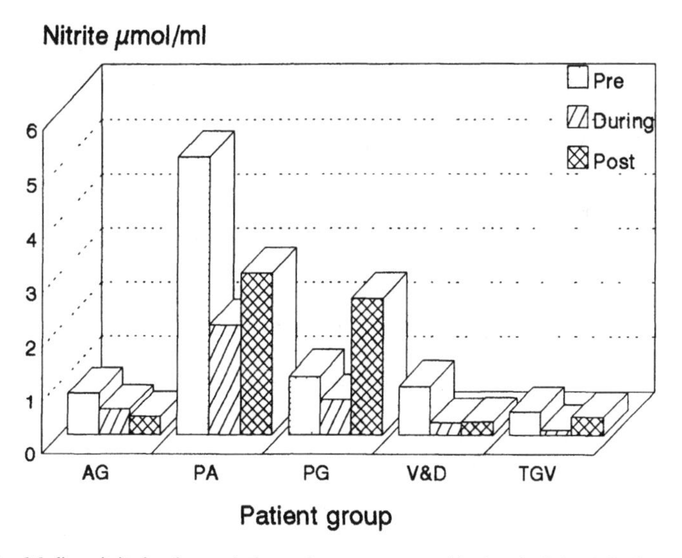

Figure 4. Median nitrite levels pre, during and post treatment with vitamin C 4 g daily for 4 weeks. (AG = atrophic gastritis [n=7], PA = pernicious anaemia [n=13], PG = partial gastrectomy [n=10], V&D = vagotomy & drainage [n=26], TGV = truncal vagotomy [n=25])

Acknowledgements

I wish to thank Belinda J Johnston for her active and constructive collaboration and help during the studies carried out in the department and quoted in the text and for preparing this manuscript for publication.

References

1. C. Muir, J. Waterhouse, T. Mack, J. Powell & S. Whelan, eds., Cancer incidence in Five Continents, Vol V, IARC Sci Publ. No 75, Lyon, IARC, (1987).
2. W. Haenszel & M. Segi, Stomach cancer among the Japanese, *in*: "UICC Monograph Series, " Vol 10, Spinger Verlag, Berlin (1967).
3. M.J. Hill. Epidemiology and mechanism of gastric carcinogenesis, *in*: "New Trends in Gastric Cancer: Background and Videosurgery,". P.I. Reed, M. Carboni, B.J. Johnston & S. Guadagni, eds., pp 3-12, Kluwer Academic Publishers, Dordrecht/Boston/London (1990).
4. T. Hirayama. Actions suggested by gastric cancer epidemiological studies in Japan. *in*: "Gastric carcinogenesis," P.I. Reed & M.J. Hill, eds., pp 209-227, Excerpta Medica, Amsterdam (1988).
5. T. Hirayama, Diet and cancer, feasibility and importance of prospective cohort study, *in*: "Diet and Human Carcinogenesis," J.V. Joosens, M.J. Hill, J. Geboers, eds., International Congress Series 685. pp 191-198, Excerpta Medica, Amsterdam (1985).
6. P. Lauren, The two histological main types of gastric carcinoma: diffuse and so-called intestinal type. *Acta Path Microbiol Scand.* 64:31-49 (1965).
7. P. Correa, W. Haenszel, C. Cuello, S. Tannenbaum & M. Archer, A model for gastric cancer epidemiology. *Lancet* ii:58-60 (1975).
8. P. Correa, The epidemiology of gastric cancer, *World J Surg*. 15:228-234 (1991).
9. H. Ohshima & H. Bartsch, Quantitative estimation of endogenous nitrosation in humans by monitoring N-nitrosoproline excreted in the urine. *Cancer Res*, 41:3658-3662 (1981).
10. P. Bogovski & S. Bogovski, Animal species in which N-nitroso compounds induce cancer. Special report. *Int J Cancer.* 27:471-474 (1981).
11. D. Schmähl & H.R. Scherf, Carcinogenic activity of N-nitroso-diethylamine in snakes *(Python reticulatus* Biodae), *in*: "N-nitroso compounds: Occurrence, Biological effects and Relevance to Human Cancer," H. Bartsch, I.K. O'Neill, R. Von Borstel, C.T. Miller, J. Long & H. Bartsch, eds., IARC Scient. Publ, 57; pp. 677-682, IARC, Lyon (1984).
12. M.J. Hill, N-nitroso compounds and human cancer, *in*: "Nitrosamines: toxicology and microbiology," M.J. Hill, ed., VCH publishers, Basle pp 142-162 (1988).
13. D. Forman, The etiology of gastric cancer, *in*: "Relevance to Human Cancer of N-nitroso compounds, tobacco smoke and mycotoxins," I.K. O'Neill, J. Chen & H. Bartsch, eds., pp 22-32 IARC Scientific Publication no 105, Lyon. (1991)
14. P.E. Hartman, Putative mutagens and carcinogens in foods. I. Nitrate/nitrite ingestion and gastric cancer mortality, *Envir Mutagens* 5:111-114 (1983).
15. D. Forman, S. Al-Dabbagh & R. Doll, Nitrates, nitrites and gastric cancer in Great Britain. *Nature* 313:620-625 (1985).
16. H.A. Risch, M. Jain, N.W. Choi, J.G. Fodor, C.J. Pfeiffer, G.R. Howe, L.W. Harrison, K.J.P. Craib & A.B. Miller, Dietary factors and the incidence of cancer of the stomach. *Am J Epidemiol.* 122:947-959 (1985).
17. S. Al-Dabbagh, D. Forman, D Bryson, I. Stratton & R. Doll, Mortality of nitrate fertiliser workers, *Br J Ind Med.* 43:507-15 (1986)
18. R. Preussmann & A.R. Tricker, Endogenous nitrosamine formation and nitrate burden in relation to gastric cancer epidemiology, *in*: "Gastric carcinogenesis," P.I. Reed & M.J. Hill, eds., pp 147-162, Excerpta Medica, Amsterdam (1988).
19. G.W. Xu, Gastric Cancer in China: a review. *J Roy Soc Med.* 74:210-211 (1981).
20. G. Xu, P. Song & P.I. Reed, The relationship between gastric mucosal changes and nitrate intake via drinking water in a high-risk population for gastric cancer in Moping country, China, *Eur J Cancer Prevent.* 1:437-444 (1992).
21. P.I. Reed, Diet and precancerous lesions of the stomach, *in*: "Causation and prevention of human cancer," M.J. Hill & A. Giacosa, eds., pp 49-68, Kluwer Academic Publishers, Dordrecht/ Boston/London, (1990).

22. N. Dungal, The special problem of stomach cancer in Iceland: with particular reference to dietary factors, *JAMA*. 178:789-798 (1961).
23. D.R. Hitchcock & S.L. Scheiner, The early diagnosis of gastric cancer, *Surg Gynecol Obstet* 113:665-672 (1965).
24. E. Bjelke, Epidemiologic studies of cancer of the stomach, colon and rectum: with special emphasis on the role of diet, *Scand J Gastroenterol*. 9: (Suppl 31) 1-235 (1974).
25. P. Correa, C. Cuello & L.F. Fajardo, Diet and gastric cancer: Nutrition survey in a high-risk area, *J Natl Cancer Inst*. 70:673-678 (1983).
26. J.H. Weisburger, H. Marquardt & H. Hirota, Induction of cancer of the glandular stomach by extract of nitrite treated fish. *J Natl Cancer Inst*, 64:163-167(1980).
27. K. Wakabayashi, M. Nagao, T.H. Chung, M. Yin, I. Karai, M. Ochiai, T. Tahira & T. Sugimura, Appearance of direct-acting mutagenicity of various foodstuffs produced in Japan and Southeast Asia on nitrite treatment, *Mutat Res*. 158: 19-124 (1985).
28. D. Yang, S.R. Tannenbaum, B. Bucki & G.C.M. Lee. 4-chloro-6-methyoxyindole is a precursor of a potent mutagen that forms during nitrosation of the fava beans (Vicia faba), *Carcinogenesis* 5:1219-1224 (1984).
29. K. Wakabayashi, M. Nagao, M. Ochiai, T. Tahira, Z. Yamaizumi, & T. Sugimura, A mutagen precursor in chinese cabbage, indole-3-acetonitrile which becomes mutagenic on nitrite treatment, *Mutat Res*. 143:17-21 (1986).
30. K.W. Huber & W.K. Lutz, Methylation of DNA by incubation with methylamine and nitrite. *Carcinogenesis* 5:1729-1732 (1984).
31. M. Ochiai, K. Wakabayashi, M. Nagao & T. Sugimura, Tyramine is a major mutagen precursor in soy sauce, being convertible to a mutagen by nitrite, *Gann*. 75:1-3 (1984).
32. T. Hirayama, Changing Patterns in the Incidence of Gastric Cancer, *in*, "Gastric Cancer, "(1980)
33. P.C. Greus, C.C. Vizcaino, J.L.A. Sanchez, D.C. Piquer, & S.T. Serrulla, Levels of household refrigeration of foodstuffs and mortality through stomach cancer in Spain (1960-86). *Europ J Cancer Prevent*, 1: (Suppl 1) 25-26(1991).
34. J.V. Joosens & H. Kesteloot. Diet and stomach cancer, *in*: "Gastric carcinogenesis," P.I. Reed & M.J. Hill, eds., pp 105-126, Excerpta Medica, Amsterdam (1988).
35. J.V. Joosens & J. Geboers, Nutrition and gastric cancer, *Nutr Canc*. 2:250-261 (1981).
36. S. Graham, W. Schotz & P. Martino, Alimentary factors in the epidemiology of gastric cancer. *Cancer* 30:927-938 (1972).
37. J. Hu, S. Shang, E. Jia, Q. Wang, Y. Liu, Y. Wu & Y. Cheng, Diet and cancer of the stomach: a case-control study in China. *Int J Cancer* 41:331-335 (1988).
38. W. Jedrychowski, J. Wahrendorf, T. Popiela & J. Rachtan, A case-control study of dietary factors and stomach cancer risk in Poland. *Int J Cancer* 37:837-842 (1986).
39. C. La Vecchia, E. Negri, A. Decarli, B. D'Avanzo & S. Franceschi, A case-control study of diet and gastric cancer in Northern Italy. *Int J Cancer* 40:484-489 (1987).
40. H.A. Risch, M. Jain, N.W. Choi, J.G. Fodor, C.J. Pfeiffer, G.R. Howe, L.W. Harrison, K.J.P. Craib & A.B. Miller, Dietary factors and the incidence of cancer of the stomach. *Am J Epidemiol*. 122:947-959 (1985).
41. D. Trichopoulos, G. Ouranos, N.E. Day, A. Tzonou, O. Manousos, C. Papadimitriou & A. Trichopoulos, Diet and cancer of stomach: a case-control study in Greece. *Int J Cancer* 36:291-297 (1985).
42. E.D. Acheson & R. Doll, Dietary factors in carcinoma of the stomach: a study of 100 cases and 200 controls, *Gut* 51:126-131 (1964).
43. K.F. Gey, G.B. Brubacher & H.B. Stähelin, Plasma levels of antioxidant vitamins in relation to ischaemic heart disease and cancer. *Amer J Clin Nutr*. 45: (Suppl 5) 1368-1377 (1987).
44. M. Eichholzer-Helbling & H.B. Stähelin, Plasma antioxidant vitamins and cancer risk: 12-year follow up of the Basle Prospective Study. *Int J Vit Nutr Res*. 61:271 (1991).

45. C.E. West & W.A. Van Staveren, ECP-EURONUT intestinal metaplasia study: Design of the study with special reference to the development and validation of the questionnaire, *in*: "Gastric carcinogenesis," P.I. Reed & M.J. Hill, eds., pp 105-126, Excerpta Medica, Amsterdam (1988).
46. P.I. Reed, N-nitroso compounds, gastric carcinogenesis and chemoprevention, *in*: "New Trends in Gastric Cancer: Background and Videosurgery,". P.I. Reed, M. Carboni, B.J. Johnston & S. Guadagni, eds., pp 21-30 Kluwer Academic Publishers, Dordrecht/Boston/London (1990).
47. UK Subgroup of the ECP-EURONUT IM Study Group, Plasma vitamin levels in patients with intestinal metaplasia and in controls. *Eur J Cancer Prevent.* 1:177-186 (1992).
48. F. Megraud, Epidemiology of *Helicobacter pylori* Infection, *in*: "*Helicobacter pylori* and Gastroduodenal Disease," B.J. Rathbone & R.V. Heatley, eds., pp 107-123 Blackwell Scientific Publications, Oxford (1992).
49. P. Correa & B. Ruis, *Helicobacter pylori* and Gastric Cancer, *in*: "*Helicobacter pylori* and Gastroduodenal Disease," B.J. Rathbone & R.V. Heatley, eds., pp 158-164 Blackwell Scientific Publications, Oxford (1992).
50. T. Rokkas, G. Popotheodorou & N. Kaldgeropoulos, *H-pylori* infection and gastric juice ascorbic acid levels. *Irish Med J.* 161: Suppl 10, 24 (1992).
51. G.M. Sobala, C.J. Schorah, S. Shires & A.T.R. Axon, Impairment of gastric ascorbic acid concentration by acute *Helicobacter pylori* infection. *Gut* 31: A1180 (1990).
52. G.M. Sobala, C.J. Schorah, S. Shires, D.A.F. Lynch, B. Gallacher, M.F. Dixon & A.T.R. Axon, Increase in gastric juice ascorbic acid concentrations after eradication of *Helicobacter pylori*: A preliminary report. *Gut* 33; Suppl 2 S60 (1992).
53. P.I. Reed, B.J. Johnston, C.L. Walters & M.J. Hill, Effect of ascorbic acid on the intragastric environment in patients at increased risk of developing gastric cancer. *Gastroenterol.* 96:A411 (1989).
54. S. Krytopoulos, Nitrosamines in the environment. Health dangers and intervention possibilities, *in*: "Environmental Carcinogens. The problem in Greece," Proc Panhellenic Congress of Greek Society of Preventive Medicine, March 1984. (1984).

DIET, COELIAC DISEASE AND GASTROINTESTINAL NEOPLASM

Gabriele Mazzacca

Chair of Gastroenterology and Endoscopy
Universita' Federico II
Via S. Pansini, 5
Napoli, Italy

INTRODUCTION

The epidemiology of coeliac disease is likely underestimated. Coeliac disease is becoming increasingly recognized in patients presenting only isolated iron, folate or calcium deficiencies (1,2) and in asymptomatic relatives of coeliac patients (3), so called latent coeliac disease. As consequence the malignancy risk claimed to be linked to gluten intolerance and villous atrophy may be underestimated.

ADULT COELIAC DISEASE AND MALIGNANCY

Lymphoma complicating coeliac disease was first described by Gough (4). Since then, the association of lymphoma to coeliac disease has been recognized to be not uncommon. Patients with coeliac disease have an estimated risk of developing lymphoma 50 to 100 fold greater than the general population (5). More generally the risk of developing digestive neoplasia other than lymphoma is increased 5 to 10 times compared to the general population (6). Malignancy in coeliac patients is more common in males and in the elderly (6).

Lymphoma in patients with a flat jejunal biopsy may present itself in two main ways. First, clinical deterioration may arise in coeliac patients previously well controlled on a gluten-free diet, or secondly the development of symptoms or sudden worsening of longstanding moderate symptoms allow the diagnosis of coeliac disease and lymphoma to be made more or less at the same time. (4, 6).

In this latter circumstance patients show no response or only a transient response to gluten- free diet. Do these patients have coeliac disease? There is some disagreement about this point and some authors consider intestinal lymphoma responsible *ab initio* for the entire clinicopathological syndrome (7).

According to O'Farrelly et al (8) the enteropathy associated lymphoma is different from uncomplicated coeliac disease with respect to alphagliadin antibody status and sex incidence. Patients with enteropathy associated lymphoma would be male, alphagliadin

Advances in Nutrition and Cancer, Edited by
V. Zappia *et al.*, Plenum Press, New York, 1993

antibody negative and would respond poorly to gluten withdrawal. However O'Farrelly's findings are not in contrast with the possibility that lymphoma may complicate previously subclinical coeliac disease determining the loss of the ability either to mount the antibody response or to respond well to gluten-free diet. More recently Wright et al have described a 59 year-old man with intestinal malabsorption who underwent a partial jejunal resection (7). Immunocytochemical data and patterns of DNA rearrangements led to the diagnosis of T cell lymphoma. Chemotherapy was useless. On the other hand the patient showed a good response to gluten-free diet. The Authors concluded that adult coeliac disease may be a low grade lymphoma of intraepithelial T lymphocytes which is analogous to mycosis fungoides. Again the possibility that slow progressive T cell lymphoma may complicate latent coeliac disease does not seem to be ruled out.

Many data support the view that most, if not all the patients presenting with subtotal villous atrophy and lymphoma have coeliac disease.

Villous atrophy is recognized in 10% of asymptomatic first degree relatives of coeliac patients (2). On the other hand relatives of patients with lymphoma have been described to have celiac disease (9).

A case reported by Freeman and Chin (10) clearly indicates that latent coeliac disease may be present in patients who become clinically symptomatic because of the further development of lymphoma.

Most patients with subtotal villous atrophy and lymphoma do not respond to gluten withdrawal. In some cases however clinical and morphological response to gluten-free diet occurred after chemiotherapy for lymphoma (5, 11).

Finally it is well known that a number of events , such as a pregnancy, intercurrent illness or surgery may precipitate symptoms in previously asymptomatic subjects allowing the diagnosis of coeliac disease to be made. Therefore it is not unreasonable to suppose that the development of lymphoma is yet another event which can precipitate symptoms and lead to the diagnosis of coeliac disease.

Although patients with coeliac disease may develop Hodgkin's lymphoma, this is very rare (5). The large majority of intestinal lymphoma appears to be a T- cell lymphoma.

FOLLOW-UP

The risk of lymphoma and more generally of neoplasm provides a cogent reason for lifelong observation of coeliac patients.

Many clinical studies have been undertaken to assess the existence of markers consenting the follow-up of coeliac patients, in particular for those who showed very few symptoms and focal hematological tests alteration.

L.D'Agostino et al have found that postheparin plasma diamine oxidase is a helpful marker of the status of small bowel mucosa. Diamine oxidase is an enzyme whose low plasma values are enhanced by an intravenous injection of heparin which releases the enzyme from the enterocytes villous tip.

Plasma diamine oxidase values are decreased in subjects with a malabsoption syndrome, in particular in patients with active coeliac disease and rise together with the improved intestinal absorption functions after the beginning of gluten-free diet (12).

D'Argenio et al (13) have demonstrated that serum transglutaminase, an active form of factor XIII involved in mucosal repairing process, is similarly related to the activity of the disease. In this study transglutaminase activity has been found to be increased in mucosa of active coeliac disease patients and at the same time decreased in serum. . In coeliac patients on gluten-free diet serum transglutaminase levels return within normal values while mucosal levels significantly decrease.

Prospective studies are needed to evaluate the usefulness of these tests in follow-up of the coeliac patient expecially with regard of the onset of malignancy.

Recently Moroz (14) showed that a specific placental ferritin isoform was not detectable in serum of healthy blood donors but it was raised in patients with lymphoproliferative disorders. Dinari et al (15) found that serum placental ferritin values are increased in patients with active coeliac disease. The authors hypothesized that since placental ferritin is an immunosuppressive agent, this may be one of the necessary steps in the development of malignancy associated with coeliac disease.

It is not clearly known if strict adherence to gluten-free diet will reduce the risk of malignancy. However at least two reports suggest that gluten-free diet may prevent malignancy complications (16, 17).

Theoretically, the more mitoses occur, the greater the chance of mutant cells arising (18), and gluten-free diet may protect coeliac patients from malignancy by reducing inflammation and cell turnover.

Current concepts of cancer view the development of malignancy as a multistage process initiated in most cases by enviromental carcinogens. The mucosal damage in coeliac disease may increase the permeability of the intestine to carcinogens, perhaps even gluten itself. The immunological disturbances, in addition, may increase the susceptibility of coeliac patient to oncoviruses, thus triggering the appearance of malignancy.

All things considered, a gluten-restricted diet seems to be the only choice for coeliac patients. Switching from a wheat-based diet to a gluten-free diet is a challenge for adult patients. It is not surprising then that their compliance to the diet is quite low. It has been shown that a minimal dose of gluten, 1-5 mg/die can cause villous damage and lymphoid cell hyperplasia in the mucosa of coeliac patients (18). The minor compliance may simply explain the lack of responsiveness of adult coeliac patients to gluten-free diet. Nevertheless the patients should be encouraged to mantain a gluten-free diet for life.

REFERENCES

1. J.E. McGuigan, W. Volwiler. Celiac sprue: malabsorption of iron in the absence of steatorrhea. *Gastroenterology* 47:636 (1964).
2. A.J. Moss, C. Waterhouse, R. Terry. Gluten-sensitive entheropathy with ostemalacia but without steatorrhea. *N Engl. J. Med* 272:825 (1965).
3. W.C. MacDonald, W.O.Dobbins, C.E. Rubin. Studies of the familiar nature of coeliac sprue using biopsy of small intestine. *N.Engl J.Med* 272:448 (1965).
4. K.R. Gough, A.E. Read, J.M. Naish. Intestinal reticulosis as a complication of idiopathic steatorrhoea *Gut* 3:232 (1962)
5. G.K.T. Holmes, P.L. Stokes, T.M. Sorahan, P. Prior, J.A.M. Waterhouse, W.T. Cooke. Coeliac disease, gluten-free diet and malignancy. *Gut* 17:612 (1976).
6. W.S. Selby, N.D. Gallaher. Malignancy in a 19 year experience of adult celiac disease. *Dig. Dis. Sci.* 24:684 (1979).

6 C.M. Swinson, G. Slavin, E.C. Coles, C.C. Booth. Coeliac disease and malignancy. *Lancet* i:111 (1983).

7. D.H. Wright, D.B. Jones, H. Clark, G.M. Mead, E. Hodges, Howell W.M. Is adult coeliac disease due to a low-grade lymphoma of intraepithelial T lymphocytes? *Lancet* 337: 1373 (1991).
8. F. O'Farrelly,C.Feighery, D.S.O'Briain,et al. Humoral response to wheat protein in patients with coeliac disease and enteropathy associated T-cell lymphoma. *Br Med J.* 293: 908 (1986).
9. B.T. Cooper, G.F.J Holmes, R. Ferguson, W.T. Cooke. Coeliac disease and malignancy. *Medicine* (Baltimore) 59:249 (1980).

10 H.J.Freeman and B.K. Chiu, Multifocal small bowel lymphoma and latent celiac sprue. *Gastroenterology* 90:1992 (1986).

11. C. O'Farrelly, C. Feighery, D.S. O'Brian et al Humoral response to wheat protein in patients with coeliac disease and entheropathy associated T-cell lymphoma. *Br. Med. J.* 293:908 (1986).

12. L.D'Agostino, C.Ciacci, B.Daniele, M.V.Barone, R.Sollazzo, G.Mazzacca. Postheparin plasma diamine oxidase in subjects with small bowel mucosa atrophy. *Dig. Dis. Sci* 34: 3 313 (1987)
13. G.D'Argenio, I. Sorrentini, C.Ciacci, S.Spagnuolo, R. Ventriglia, A. De Chiara, G. Mazzacca. Human serum transglutaminase and coeliac disease: correlation between serum and mucosal activity in an experimental model of rat small bowel enteropathy. *Gut* 30 7:950 (1989).
14. Moroz C, Bessler H, Lurie Y, Shaklai M New monoclonal antibody enzimoassay for the specific measurement of placental ferritin isotype in hematologic malignancies. *Exp. Hematol* 15:258 (1987).
15. Dinari G, Zahavi I, Marcus H, Moroz C. Placental ferritin in coeliac disease: relation to clinical stage, origin, and possible role in the pathogenesis of malignancy. Gut 32 999 (1991).
16. R.E. Barry and A.E. Coeliac disease and malignancy Quarterly J. of Medicine 42:665 (1973).
17. G.K.T. Holmes, P.Prior, M.R. Lane, D.Pope, R.N.Allan. Malignancy in coeliac disease- effect of a glutn free diet. Gut 30: 330 (1989).
18. W.J. Austad, J.S.Cornes, K.R. Gough C.F. McCarthy, A.E. Read. Steatorrhoea and malignant lymphoma. The relationship of malignant tumours of the lymphoid tissue and coeliac disease. Am J. Dig. Dis. 12:475 (1967).
19. P.J. Ciclitira, H.J. Ellis, D.J. Evans, E.S. Lennox. A radioimmunoassay for wheat gliadin to assess the siutability of gluten-free foods for patients with coeliac disease. Clin. Exp. Immunol: mar. 59 3:703 (1985).

PROTECTIVE ASPECTS OF THE MEDITERRANEAN DIET

Anna Ferro-Luzzi, Andrea Ghiselli

National Institute of Nutrition
Via Ardeatina 546
Rome, Italy

INTRODUCTION

When the concept of Mediterranean diet, with its health promoting connotations, emerged thanks to Ancel Keys[1], it was in relation to the risk of ischemic heart disease. The lipid hypothesis was the prevalent and the attention was focused on the fat moiety of the diet.

Since then a large body of evidence has been accumulated on the role of reactive oxygen species (ROS) in the initiation as well in the promotion of multistage carcinogenesis. At the same time it became apparent that many nutrient and non-nutrient compounds are involved in the metabolism of ROS, thus suggesting a nutritional modulation of the risk of cancer. Several nutrients appear to play key roles in cancer prevention or promotion. For example epidemiological and experimental researches have suggested a link between high fat intake and some types of cancer (breast, large bowel and prostate)[2], and the reduction of caloric intake appears capable of reducing tumour incidence in animal models[3].

A large inter-country variability of cancer rates exists in Europe, with a tendency for the frequency of various cancers to be lower in Mediterranean Countries than in Northern Europe[4]. The lower cancer risks of the Mediterranean people are now being linked to the prevalent dietary typology, low in saturated fatty acids and high in fibre, starch, and antioxidant vitamins.

While the health benefits of the Mediterranean dietary habits are fairly well recognised, the exact definition of what this diet consists of is not straightforward, and over the recent years many diverse attempts have been made to describe it[5]. It appears that there may be more than one "variation on the theme", and that the "theme" may not be defined on a purely geographic basis. In fact the comparison of the national diets of the various Mediterranean Countries reveals that the difference among them is almost as large as that between the Mediterranean Countries and the other European Countries[6]. In a previous paper we defined the typical Mediterranean diet as being that consumed by the rural population of Southern Italy in the early '60s[6].

Since then many changes have taken place in the dietary style of Southern Italy, and the differences from Northern Italy have currently become greatly attenuated. It appears however

Advances in Nutrition and Cancer, Edited by
V. Zappia *et al.*, Plenum Press, New York, 1993

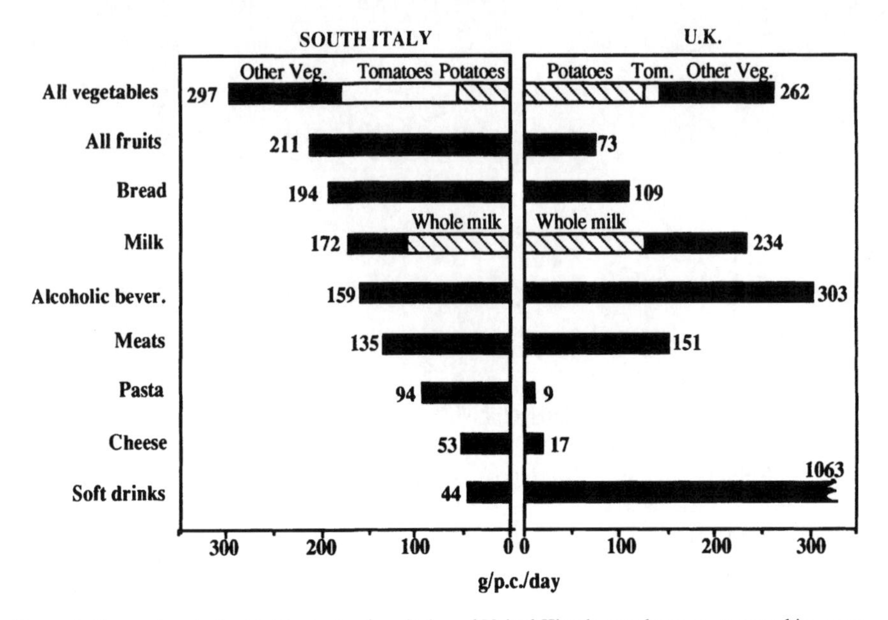

Figure 1. Comparison in food intake in Southern Italy and United Kingdom; values are expressed in grams pro capite per day. - From A. Ferro-Luzzi et al. in preparation.

that some differences still exist, which become particularly evident when the current dietary profile of South Italy[7] is compared to that of Northern European Countries. The example reported in Figure 1 illustrates the comparison with Great Britain[8]. It shows that the British consume remarkably less fruits, bread, pasta and cheese, and more milk, alcoholic and non-alcoholic beverages. Moreover, in spite of similar intakes of vegetables, there are profound differences in the type of vegetables, the Italians eating more tomatoes and the British more potatoes.

OXIDATIVE STRESS AND DISEASE

In the 80's it became apparent that our understanding of the pathogenic processes of atherosclerosis was incomplete and experimental evidence showed that the lipid hypothesis was incapable of fully accounting for the disease. The antioxidant hypothesis[9] was included among the causative factors linking diet to cancer promotion as well as to atherosclerosis, and the oxidative damage of susceptible substrates was identified as a mechanism responsible for the early atherosclerotic lesion and for the initiation of neoplastic processes.

In vitro experiments have firmly shown that the hydroperoxide produced by the attack of aggressive forms of oxygen acting upon a suitable substrate, enhances the uptake of low density lipoproteins (LDL) by macrophages and drives the formation of lipid-laden foam cells in the arterial intima[10]. For what concerns cancer disease it is recognised that DNA can undergo ROS attack, resulting in the addition of hydroxyl radical to pyrimidine base's double bonds and the abstraction of hydrogen from the sugar moiety. The subsequent chain scission can cause cell mutagenesis and carcinogenesis[11,12].

NATURAL ANTIOXIDANTS AND CANCER

The human organism is equipped with an antioxidant system that protects it from the hazard deriving from ubiquitous ROS Various competent enzymes and other compounds endowed with antioxidant capacity represent a formidable defence system[13].

The link between diet and cancer might be seen as resulting from an imbalance in the intake of oxidants and antioxidants. The evidence to support a relationship between the occurrence of cancer and the diet is good for specific foods[14], slightly less good for individual nutrients[15]. In an excellent meta-analysis of 156 selected studies relating cancer death rates to plant material in the diet, Block[16] found that individuals with the lowest quartile of vegetable and fruit intake had about twice the risk of dying from cancer than the highest quartile of the distribution. There seemed to be a graded risk over the whole range of intakes with a negative dose-response type relationship observed in 82% of all the studies. A summary of the review is reported in Figure 2.

The existence of the relationship between fruit and vegetable intake and cancer risk has been explained mainly by the presence of antioxidant vitamins and pro-vitamins in foods of vegetal origin, especially vitamins C, E, and A and its pro-vitamin β-carotene These compounds exercise a well recognised antioxidant action, either acting as ROS scavengers or as chain-breakers[17,18].

More recently the attention of epidemiologists and nutritionists has been attracted by other compounds present in our foods which have no known nutrient attribute, but may play an important role in protecting from oxidative damage[19] Up to now these non-nutrient compounds have elicited little attention, mainly from food technologists who study them because of the various properties they impart to foods of vegetal origin. Thus for example the vivacious colours of many fruits, vegetables and red wines derive from their anthocyanine content[20], organic sulphides such as diallylsulphide impart the fragrance -- or foul odour,

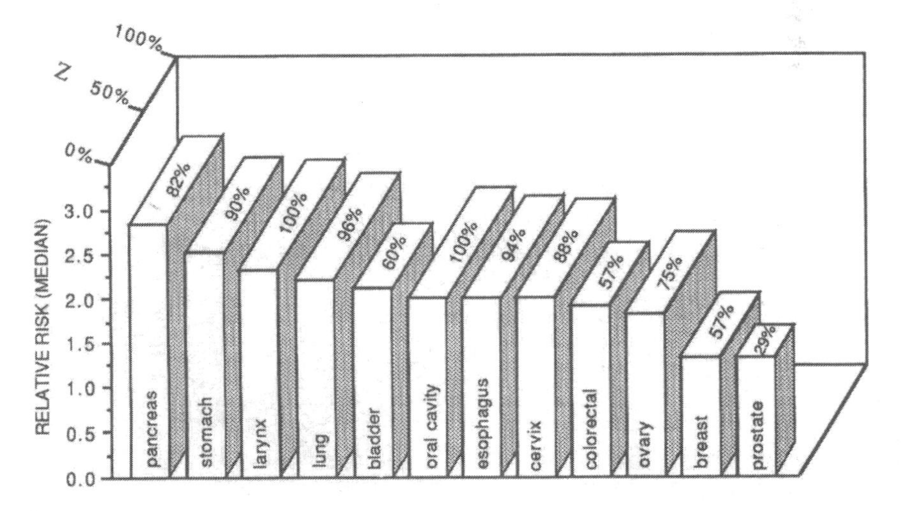

Figure 2 Summary of epidemiological studies on fruit and vegetable intake and cancer risks, values are expressed as increase in cancer death risk of people in the first quartile of fruit and vegetable intake, with respect to the fourth quartile people In the axis "Z" the percentage of the studies finding inverse correlation between fruit and vegetable intake and cancer incidence is reported Data source G Block et al [16].

depending on the point of view -- to the Mediterranean diets abundantly spiced with garlic[21]; the balance between various polyphenols in wines and olive oil plays an important role in the perceived taste, etc.

There is now mounting evidence that these non-nutrient substances are not metabolically inert, and it appears that they can exert a more or less powerful antioxidant action. Thus for example, lycopene has been found to be the most efficient carotenoid in quenching singlet oxygen[22]. Canthaxanthin, astaxanthin and hydroxyl-containing xantophylles are more powerful than β-carotene in retarding the formation of hydroperoxide in a peroxyl radical-mediated lipid peroxidation[23]. Glucosinolates and dithiothiones, present in cruciferous vegetables, have been shown to protect from colon cancer, possibly through their ability to stimulate the detoxication of carcinogens[24].

Other compounds, such as the organic sulphides found in garlic and onions, appear to have protective action against stomach cancer, possibly related to their antagonism towards alkylating carcinogens[21]. It is interesting to point out that the oily extract of garlic and onion inhibits platelet aggregation *in vivo* and *in vitro*, by blocking thromboxane synthetase[26]. These oils have been shown to exert also fibrinolytic, hypotensive, hypoglycemic, or hypocholesterolemic actions (see Kendler[27] for a review).

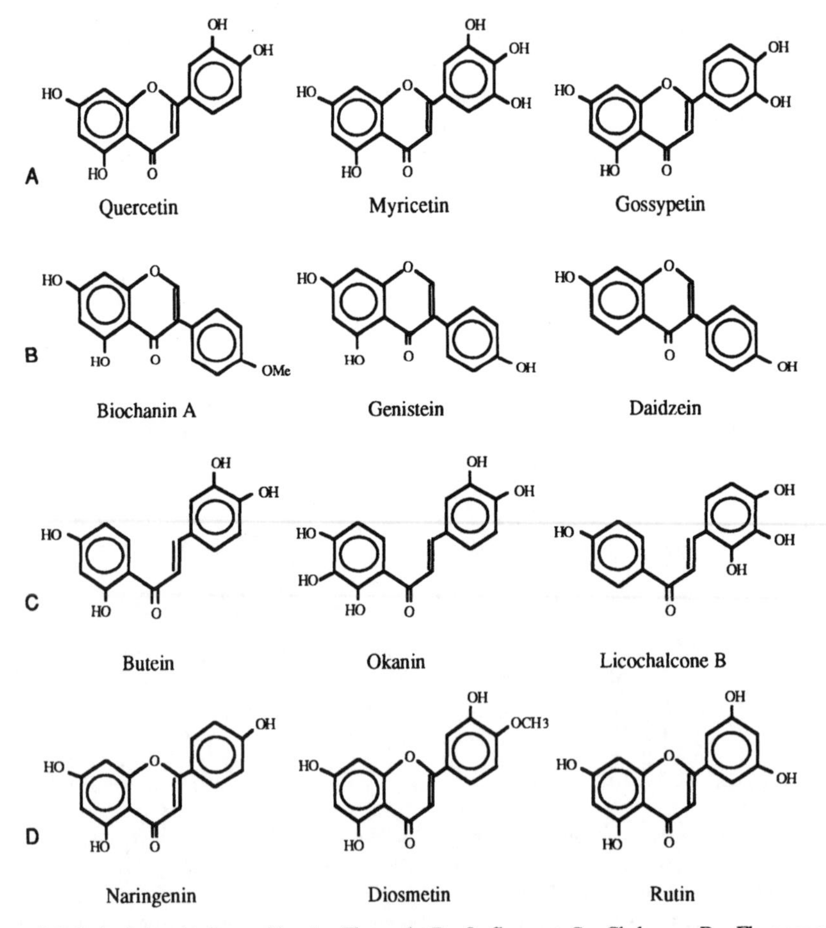

Figure 3. Principal plant bioflavonoids : A = Flavonols; B = Isoflavones; C = Chalcones; D = Flavonones. Modified from Namiki[25].

Another very interesting compound is the omni-present ubiquinone, which might turn out to be one of the most effective and powerful *in vivo* lipid antioxidants[28].

A very large class of compounds is emerging as one of potentially crucial importance. These are the polyphenols, a large family of compounds practically ubiquitous in all plant material[20]. Some of the simple phenols, such as hydroxytyrosol, have an antioxidant activity greater than that of buthylated hydroxytoluene, a powerful synthetic antioxidant[29]. Other phenolic acids are thought to prevent interactions of benzo(a)pyrene with DNA[30].

The family of polyphenols includes bioflavonoids (Figure 3), anthocyanines, procyanidines and tannins and counts more than 3000 known compounds, most of which still need to be investigated regarding their metabolism, absorption and bioavailability[20].

These substances may inhibit the oxidative stress at several steps as reported in Table 1. The non enzyme-mediated lipid peroxidation leads to peroxide production with consequent damage and disruption of cellular membranes, DNA oxidation and cellular death. Neighbouring cells would be stimulated to proliferate and this could represent a promotional stimulus for cancerogenesis[31]. Polyphenols are capable of inhibiting this phenomenon, by scavenging free radicals and oxidising agents and by breaking peroxidative chain reactions[32].

Table 1. Main activities of bioflavonoids on oxidative stress

DIRECT ACTIONS	MEDIATED ACTIONS
1. Inhibition of initiation of lipid peroxidation.	1. Inhibition of phospholipase A2
2. Inhibition of chain propagation.	2. Inhibition of lipooxigenases
	3. Inhibition of phosphodiesterase.

The indirect, mediated, effect of polyphenols may consist in the inhibition of several possible steps of relevant enzymic pathways; more specifically they could prevent, through their inhibition of phospholipase A2[33], the release of arachidonic acid from membrane phospholipids, thus inhibiting the pathway leading to the production of chemiotactic and inflammatory substances[34]. Furthermore by inhibiting the lipooxygenase pathway[33], polyphenols block hydroperoxyeicosatetraenoic acid (HPETE) and leukotriene production, thus eliminating a powerful inhibitor of prostacyclin synthetase and chemiotactic substances[34]. Polyphenols are also capable of inhibiting phosphodiesterases with subsequent increase in intracellular cAMP[35]. cAMP levels of platelets modulate platelet responsiveness to aggregating agents, high levels inhibiting and low levels stimulating platelet function[35].

The promotion of a non-reactive status of the platelets appears particularly important also in the light of recent results suggesting that platelets may produce hydroxyl radicals during aggregation[36]. This production of hydroxyl radicals would initiate or aggravate the oxidative stress status and the oxidative damage to cellular components. Platelets may therefore play an important role in carcinogenesis, not only in tumour progression through the release of chemiotactic, vasoconstrictor and inflammatory substances, or through the inhibition of vasodilator substances[37], but also directly through the release of ROS. It is well known that the oxidising species deriving from the cyclooxygenase pathway (not yet identified) are involved in "co-oxygenation" of benzo(a)pyrene and that this mechanism has a close relationship to lung carcinogenesis[38].

Polyphenols are present in processed vegetable commodities, such as wine and olive oil, two typical components of the Mediterranean Diet. The evidence of a protective role of olive oil in respect to degenerative diseases is growing. This property appears to be linked not only to its blood lipid-lowering action[39], but also to its antioxidant action. The major components of olive oil are essentially oleic acid (75%), a monounsaturated fatty acid, linoleic acid, tocopherols (almost exclusively α-tocopherol), and β-carotene. But there are also minor components, represented by phenolic compounds, sterols and terpenes which could be involved in the modulation of oxidative stress *in vivo*[40]. A recent work has shown that plasma and lipoprotein (VLDL+LDL) content of thiobarbituric acid-reactive substances (TBA-RS) and conjugated dienes is significantly higher in rats fed either soya oil or trioleine (a mixture containing 75% of tryglicerides of oleic acid) than in olive oil-fed rats[41]. Furthermore, the olive oil-fed animals have a higher antioxidant status measured as TRAP (Total Radical-trapping Antioxidant Parameter). These effects are not accounted by vitamin E, because all rats received the same amount of α-tocopherol[41].

CONCLUDING REMARKS

Considering the potentially critical role of these non-nutrients it is most important to know more about their presence in terms of quantity and quality in the Mediterranean diet. Unhappily this question is destined to remain unanswered in the near future. Although work is now being conducted in this sector, polyphenols and other non-nutrients are not yet listed along with nutrients in tables of food composition and it is impossible to calculate the intake.

It must be recognised that the measurement of these minor compounds presents many difficulties. Problems are encountered at the stage of sampling, extraction, saponification, identification and quantification of the compounds present in complex mixtures, and the risk of artefacts and of 'wrong' answers is very high[42]. Furthermore there is also very little knowledge on polyphenols' bioavailability in humans and on their metabolic destiny.

Antioxidant compounds may be important not only in providing antioxidant substances to the body, but also in limiting the production of toxic oxidation compounds during the industrial processing of foods or their home-based preparation. Several lipid and non lipid products of food processing have been shown to be mutagenic and cancerogenic in animal models[43]. In human nutrition, although a high ingestion of oxidised food is unlikely, because of their unpalatability (rancid taste), the possibility that *in vivo* metabolic interactions of oxidised foods with vitamins, cytochromes P-450, platelet activity and lipoprotein metabolism may occur should not be neglected.

A diet rich in fresh plant material would thus exert a beneficial health-promoting action through three paths. It would cause a reduction in the intake of preformed hydroperoxides and carcinogens derived from food cooking, cooked foods being displaced by fresh fruit and vegetables. It would provide abundant antioxidants which would maximise the *in vivo* antioxidant potential. Its high fibre content would contribute to the intestinal scavenging of preformed carcinogens in the diet[44].

Thus, the mechanisms exist whereby a plant-rich diet, such as the celebrated Mediterranean diet, should be not only non-toxic as mentioned by James[45], but also actively protective and health promoting.

REFERENCES

1. A. Keys and M. Keys, "How to Eat Well and Stay Well, The Mediterranean Way", Doubleday & Co. Inc., New York; (1975).
2. K.K. Carroll, L.M. Braden, J.A. Bell and R. Kalamegham, Fat and cancer, *Cancer* 58:1818-1825 (1986).
3. D. Kritchevsky, M.K. Weber, D.M. Klurfeld, Dietary fat versus caloric content in initiation and promotion of 7,12-dimethylbenz(a)-anthracene-induced mammary carcinogens in rats. *Cancer Res.* 44:3174-3177 (1984).
4. C. Muir, J. Waterhouse, T. Mack, J. Powell, S. Whelan, M. Smans and F. Casset, "Cancer Incidence in Five Continents", Vol V. IARC, Lyon (1987).

5. F. Fidanza, The Mediterranean Italian diet: keys to the contemporary thinking. *Proc Nutr. Soc.* 50: 519-526 (1991).

6. A. Ferro-Luzzi, S. Sette, The Mediterranean Diet: an attempt to define its present and past composition. *Eur. J. Clin Nutr.* 43 (Suppl.2): 13-29 (1989).

7. A. Turrini , A. Saba , C. Lintas, Study of the Italian reference diet for monitoring food constituents and contaminants. *Nutr. Res.* 11:841-873, (1991)

8. J. Gregory, K. Foster, H. Tyler, M. Wiseman,.: "The Dietary and Nutritional Survey of British Adults". Off. Population Census, Social Survey Division, London HMSO (1990).

9. B. Halliwell, Oxidant and human disease: some new concepts. *FASEB J.* 1: 358-364, (1987).

10. T.E. Carew, D.C. Shwenke and D. Steinberg, Antiatherogenic effect of probucol unrelated to its hypocholesterolemic effect: Evidence that antioxidant *in vivo* can selectively inhibit low density lipoprotein degradation in macrophage-rich fatty streaks and slow the progression of atherosclerosis in the Watanabe heritable hyperlipidemic rabbit. *Proc. Natl. Acad. Sci. USA* 84:7725-7729; (1987).

11. R. Adelman, R.L. Saul and B.N. Ames, Oxidative damage to DNA: Relation to species metabolic rate and life span. *Proc. Natl. Acad. Sci. USA* 85:2706-2708 (1988).

12. Y. Kuchino, F. Mori, H. Kasai, H. Inoue, S. Iwai, K. Miura , E. Ohtsuka and S. Nishimura, Misreading of DNA templates containing 8-hydroxydeoxyguanosine at the modified base and at adjacent residues. *Nature* 327:77-79 (1987).

13. G.G.Duthie, K.J.W Whale . and W.P.T. James, Oxidants, antioxidants and cardiovascular disease. *Nutr. Res. Rev.* 2:51-62 (1989).

14. F. Berrino, P. Muti. Mediterranean diet and cancer. *Eur. J. Clin. Nutr.* 43 (Suppl. 2): 49-55 (1989).

15. K.F. Gey, G.B. Brubacher and H.B. Stähelin, Plasma levels of antioxidant vitamins in relation to ischemic heart disease and cancer. *Am. J. Clin. Nutr.* 45:1368-1377 (1987).

16. G. Block, B. Patterson, A. Subar, Fruit vegetables and cancer prevention: a review of the epidemiologic evidence. *Nutr. Cancer* 18:1-29 (1992).

17. T. Doba, G.W. Burton and K.U. Ingold, Antioxidant and co-oxidant activity of vitamin C. The effect of vitamin C, either alone or in the presence of vitamin E or a water- soluble vitamin E analogue, upon the peroxidation of multilamellar phospholipid liposomes. *Biochim.Biophys. Acta* 835:298-303 (1985).

18. C.H. Hennekens, S.L. Mayrent and W. Willet, Vitamin A, carotenoids and retinoids. *Cancer* 58:1837-1841 (1986).

19. B.N. Ames, Dietary carcinogens and anticarcinogens. *Science* 221:1256-1264 (1983).

20. J.B. Harborne, . Nature, distribution and function of plant flavonoids.*in*: "Plant Flavonoids in Biology and Medicine: Biochemical, Pharmacological and Structure-Activity Relationships", Cody V, Middleton E and Harborne J.B. ed., Alan R. Liss, Inc New York, (1986) pages 15-24.

21. M.J. Wargovich, Diallyl sulfide, a flavour component of garlic (allium sativum) inhibits dimethylhydrazine induced colon cancer. *Carcinogenesis* 8:487-489 (1987).

22. P. Di Mascio, M.E. Murphy and H. Sies, Antioxidant defense systems: the role of carotenoids, tocopherols and thiols. *Am. J. Clin. Nutr.* Suppl. 53, 194S-200S, (1991).

23. J. Terao, Antioxidant activity of b-carotene-related carotenoids in solution. *Lipids* 24:659-661 (1989).

24. M.J. Wargovich, New dietary anticarcinogens and prevention of gastrointestinal cancer. *Dis. Colon Rectum* 31, 72-75 (1988).

25. M. Namiki, Antioxidant/antimutagens in foods. *Food Sci. Nutr.* 29:273-300 (1990).

26. J.Y. Vanderhoek, A.N. Makheja and J.M. Bailey, Inhibition of fatty acid oxygenases by onion and garlic oils. Evidence for the mechanism by which these oils inhibit platelet aggregation. *Biochem. Pharmacol.* 29:3169-3173 (1980).

27. B.S. Kendler, Garlic (Allium sativum) and onion (Allium cepa): A review of their relationship to cardiovascular disease. *Prev. Med.* 16:670-685 (1987).

28. R. Stocker, V.W. Bowry . and B. Frei, Ubiquinol-10 protects human low density lipoprotein more efficiently against lipid peroxidation than does α-tocopherol. *Proc. Natl. Acad. Sci. USA* 88:1646-1650 (1991).

29. G. Papadopoulos and D. Boskou, Antioxidant effect of natural phenols on olive oil. *J. Am. Oil Chem. Soc.* 68:669-671 (1991).

30. P. Lesca, Protective effects of ellagic acid and other plant phenols on benzo(a)pyrene neoplasia in mice. *Carcinogenesis* 4:1961-1653 (1983).
31. B.N. Ames,. Endogenous DNA damage as related to cancer and aging Mutat. Res. (1988)
32. N.P. Das and A.K. Ratty, Effects of flavonoids on induced non-enzymic lipid peroxidation, *in*: "Plant Flavonoids in Biology and Medicine: Biochemical, Pharmacological and Structure-Activity Relationships", V. Cody, E. Middleton and J.B. Harborne. ed., Alan R. Liss, Inc New York, (1986), pages 243-247.
33. A.F. Welton, L.D. Tobias, C. Fiedler-Nagy, W. Anderson, K. Hope, K. Meyers and J.W. Coffey, Effect of flavonoids on arachidonic acid metabolism. *in*: "Plant Flavonoids in Biology and Medicine: Biochemical, Pharmacological and Structure-Activity Relationships". V. Cody, E. Middleton and J.B. Harborne ed., Alan R. Liss, Inc New York, (1986), pages 231-242.
34. W.L. Smith, The eicosanoids and their biochemical mechanisms of action. *Biochem. J.* 259:315-324, (1989).
35. A. Beretz, R. Anton, J.P. Cazenave, The effects of flavonoids on cyclic nucleotide phosphodiesterases. *in*: "Plant Flavonoids in Biology and Medicine: Biochemical, Pharmacological and Structure-Activity Relationships", V. Cody, E. Middleton and J.B. Harborne ed., Alan R. Liss, Inc New York, (1986), pages 281-296.
36. A. Ghiselli, M. Serafini and A. Ferro-Luzzi, Attivazione piastrinica e stress ossidativo. 6th Natl. Cong. of Società Italiana per lo Studio dell'Aterosclerosi, Padova, Oct 29-31, 1992.
37. M.F. McCarty, An antithrombotic role for nutritional antioxidants: implications for tumor metastasis and other pathologies. *Med. Hypotheses* 19:345-357, (1986).
38. Y.S. Bakhle, Synthesis and catabolism of cyclo-oxygenase products.*Br. Med. Bull.* 39:214-218, (1983)
39. S.M. Grundy, Monounsaturated fatty acids, plasma cholesterol and coronary heart disease. *Am. J. Clin. Nutr.* 45:1168-1175, (1987).
40. G.F. Montedoro, M. Baldioli, M. Servili, I composti fenolici dell'olio di oliva e la loro importanza sensoriale, nutrizionale e merceologica. *Giorn. Ital. Nutr. Clin Prev.* 1:19-32, (1992).
41. C. Scaccini, M. Nardini, M. D'Aquino, V. Gentili, M. Di Felice and G. Tomassi, Effect of dietary oils on lipid peroxidation and on antioxidant parameters of rat plasma and lipoprotein fractions. *J. Lipid Res.* 33:627-633, (1992)
42. K.G. Scott , Observations on some of the problems associated with the analysis of carotenoids in foods by HPLC. *Food Chem*, 45:357-364, (1992).
43. S. Kubow, Routes of formation and toxic consequences of lipid oxidation products in foods. *Free Rad. Biol. Med.* 12:63-81, (1992).
44. M.A. Dreher, Dietary fiber and its physiopathological effects. *in*: "Handbook of Dietary Fiber", M. Dekker Inc. New York (1987) pages 199-279.
45. W.P.T. James, G.G. Duthie and K.W.J. Wahle, The Mediterranean diet: protective or simply non-toxic? *Eur. J. Clin. Nutr.* 43 (Suppl. 2):31-41, (1989).

THE ROLE OF "BIOELECTRICAL IMPEDANCE ANALYSIS" IN THE EVALUATION OF THE NUTRITIONAL STATUS OF CANCER PATIENTS

Giuseppe Catalano, Michele Della Vittoria Scarpati, Ferdinando De Vita, Pasquale Federico[1], Giuseppina Guarino[1], Andrea Perrelli[1], Valentina Rossi[1]

Medical Oncology; [1]Clinical Methodology
Department of Clinical and Experimental Internistic Medicine
II University of Naples
Via Pansini 5
80131 Naples, Italy

INTRODUCTION

The malnutritional status of cancer patients is due to many factors, such as inadequate food-intake, cancer-host competition, metabolic alterations of the host organism, and complications of therapies (3, 5).

The consequent weight loss causes a further weakness of the patients, a bad response to the antiblastic therapies, and a shorter survivial (1, 4).

The aim of our study has been to evaluate the nutritional status of cancer patients by a non-invasive technique, the "Bioelectrical Impedance Analysis" (BIA). This technique allows the measurement of a few indexes of the body mass composition, providing indirect data about the nutritional condition of the patients (7).

MATERIALS AND METHODS

We studied 44 cancer patients (age 37-70 years) with ECOG performance status of 0-1 sec., undergoing no current antitumoral therapy. Of the total, 18 of them were affected by gastroenteric cancer, 9 by lung cancer; 8 by breast cancer, 7 by a N-H lymphoma, 1 by leiomyosarcoma, and 1 by adrenal cancer. In addition, we studied 30 control subjects (age 35-68 years).

None of them had a concomitant pathology causing hydroelectrolytic alterations. These patients underwent "Bioelectrical Impedance Analysis" (BIA). This survey is based on the measure of electrical conductivity of tissues with high (splanchnic tissues, skin, muscles) or low (adipose tissue and bone) hydroelectric content. By this survey it is possible to evaluate

Advances in Nutrition and Cancer, Edited by
V. Zappia *et al.*, Plenum Press, New York, 1993

lean mass (LM) and fat mass (FM) approximate amount and LM/FM ratio, total body water, Extracellular mass/Intracellular mass (ECM/BCM) ratio, and exchangeble Na+/ K+ ratio.

Exchangeble Na+/K+ ratio expresses the quantitative relation between cationic intracellular and extracellular content which, in normal subjects, is higher in the intracellular compartment. The ECM/BCM ratio instead, when increased, means a dilatation of the extracellular spaces. The BIA parameters always result altered when a malnutritional condition is present.

In addition, for every patient we also measured the difference between real and ideal weight, the thickness of the main skin folds, since they represent an indirect index of the energetic reserves of the organism, and the arm circumference and muscle area for evaluating the status of muscle protein trophism.

Furthermore, we analyzed some biochemical parameters such as albumin, prealbumin, albumin/globulin ratio, transferrin and cholinesterase, in order to evaluate the status of visceral proteins; zinc, copper and selenium plasma values were also determined since oligoelement activity is essential for the function of many enzymatic cofactors (6).

Table 1. BIA paramaters.

Parameters	Cancer patients	Control subjects	p
Fat mass	26.28 +/-78.6	27.89 +/-8.14	ns
Lean mass	73.58 +/-7.95	72.46 +/-7.06	ns
Total body water	55.61 +/- 6.96	51.45 +/-7.06	ns
ECM/BCM	1.60 +/-0.32	0.95 +/-0.10	<0.05
Exchangeble Na+/K+	1.23 +/-0.33	0 .93 +/-0.23	<0.05
Fat mass/Lean mass	3.43 +/-1.86	3.43 +/-1.86	ns

RESULTS

Bioelectrical Impedance Analysis revealed an alteration of ECM/BCM ratio and of exchangeble Na+/K+ ratio. The amounts of lean and fat body mass and LM/FM ratio, instead, did not result meaningfully altered (Table 1).

The anthropometric parameters (arm circumference, skin fold thickness) did not result meaningfully altered in the patients in comparison with control subjects similar to the patients for sex and age.

The analysis of biochemical parameters showed a decrease of plasma protein concentration and of zinc and selenium serum levels, whilst copper levels resulted increased (Table 2.).

Table 2. Biochemical parameters.

Parameters	Cancer patients	Control subjects	p
Prealbumin(mg/dl)	21;57+/-6.3	23.5+/-7.3	<0.05
Transferrin(mg/dl)	224.75+/-125.3	339.5+/-84.9	<0.05
ALB/GLOB ratio	0.93+/-0.08	1.52+/-0.18	<0.001
Cholinesterase(U/L)	433+/-2429	9764+/-4461	<0.05
Copper(g/dl)	122.6+/-44.4	95.5+/-13.2	<0.05
Zinc(g/dl)	78.6+/-11.7	87.5+/-9.7	<0.05
Selenium(g/dl)	71.1+/-16.7	85.0+/-21.7	<0.05
Total cholesterol	80.9+/-16.7	180.8 +/-123	<0.001

DISCUSSION

A careful analysis of the nutritional status of cancer patients is essential before starting a therapeutic program since malnutrition can strongly reduce the benefits of a radio- or chemo-therapy (2, 3). Bioelectrical Impedance Analysis represents therefore an important step in the evaluation of this condition together with the estimation of other anthropometric and biochemical parameters.

The qualitative information supplied by this survey about body mass composition, even more than numerical values, allows the detection of even silent malnutritional conditions (7).

As shown in our data, BIA revealed the presence of a latent malnutritional status in cancer patients whose anthropometrical indexes (weight, skin folds, circumferences) were still normal. The alterations of BIA parameters, ECM/BCM ratio and exchangeble Na+/K+ ratio, resulted meaningfully correlated to the reduction of albumin/globulin ratio and of prealbumin and transferrin levels, which represents important indexes of protein nutritional status, as some authors demonstrated (1, 2, 3, 4, 5).

Our observations about oligo element serum levels confirm that in cancer patients copper levels are often increased, while zinc levels are often decreased. The deficiency of selenium observed in our cancer patients seems remarkable. Selenium is in fact supposed to have a protective antitumoral effect, as many authors affirm (6).

A malnutritional condition together with a deficiency of oligoelements in cancer patients can further worsen the immunodeficiency due to the presence of a neoplasia.(5)

Our results justify therefore the importance of analyzing the nutritional status of cancer patients and show how this is easy to achieve by the application of a non-invasive technique such as BIA.

When a malnutritional condition is revealed, it becomes essential to plan precautiously a nutritional program aimed toward correcting this status before starting any therapies.

REFERENCES

1 American College of Physicians, 1989, Parenteral nutrition in patients receiving cancer chemotherapy, Ann Int Med.110:735

2. Bozzetti F., 1989, Effects of artificial nutrition on the nutritional status of cancer patient, J Par Ent Nutr. 406:20

3. Cheblowski RT., 1991, Nutritional support of the medical oncology patient, Hematol Oncol Clin North Am,5:1

4. De Wys WD, Begg D. et al , 1980, Prognostic effects of weight loss prior to chemiotherapy in cancer patients, Am J Med.69:491

5. Eng En NG, Lowri SF, 1991, Nutritional support and cancer cachexia, Hematol Oncol Clin North Am, 5:1

6. Federico P., Giordano C. et al, 1991,Valutazione degli oligoelementi in pazienti neoplastici,Tumori.77:3

7. Sukkar S.G., Adami G.F. et al., 1989,Valutazione della composizione corporea mediante impedenziometria bioelettrica nel paziente oncologico malnutrito,Tumori.75:4

HIGH SERUM HDL-CHOLESTEROL IN PRE- AND POST-MENOPAUSAL WOMEN WITH BREAST CANCER IN SOUTHERN ITALY

Renato Borrelli[1], Gabriella del Sordo[1], Emilia De Filippo[1], Franco Contaldo[1], Valerio Parisi [2], and Gerardo Beneduce[2]

[1]Clinical Nutrition, Medical School
University of Naples Federico II, Naples, Italy
[2]National Cancer Institute (NCI)
"Fondazione G. Pascale", Naples, Italy

SUMMARY

Up until now, conflicting results have been reported on the association between serum cholesterol and risk of breast cancer in women.

In this study, the serum concentrations of cholesterol, HDL-cholesterol, triglycerides and total lipids in women with breast cancer (BC) have been compared to those of women with benign breast disease (BBD). BC women had higher serum concentration of HDL-cholesterol both in pre- and in post-menopausal age. No difference was observed in the serum concentration of total cholesterol, triglycerides and total lipids.

These findings could be explained by an increased estrogen activity which is believed to be involved in the development of breast cancer, and in the modulation of lipid metabolism (lowering LDL-cholesterol and increasing HDL-cholesterol). High serum HDL-cholesterol could be a biochemical index of increased risk of having breast cancer.

INTRODUCTION

There are so far conflicting results on the association between serum cholesterol and breast cancer risk in women. In fact, several studies have reported an inverse association between these two conditions [1-4], while others a direct association[5-7], and finally, others yet have failed to show any significant relationship [8-10].

It is known that estrogen activity, which is believed to be involved in the development of breast cancer[11-13], modulates lipid metabolism [14-16].

In this study, the serum concentrations of cholesterol, HDL-cholesterol, triglycerides and total lipids in women with breast cancer (BC) at time of the diagnosis have been compared to those in women with benign breast disease (BBD) in pre- and post-menopausal age.

Advances in Nutrition and Cancer, Edited by
V. Zappia *et al.*, Plenum Press, New York, 1993

SUBJECTS AND METHODS

All women consecutively admitted for suspected primary breast cancer to the National Cancer Institute "Fondazione G. Pascale" in Naples from October 1990 to January 1991 were considered eligible for the study.

Sixty-six women participated in the study. No women had reported weight loss in the three months before the study. They underwent breast biopsy within a few days of the first observation.

The clinical procedure included standardized measurements of anthropometric variables (height, weight, etc) and a questionnaire designed to obtain information on smoking habits, dietary habits and life style.

Forty-two women were identified as breast cancer cases, and twenty-four as benign breast disease cases (mastitis, fibrocystic disease, etc) on the basis of breast biopsy.

A fasting blood sample was also taken to be analyzed for total serum cholesterol, HDL-cholesterol, triglycerides and total lipids. The serum cholesterol, HDL-cholesterol, triglycerides and total lipids were measured according to the method used in the Lipid Research Clinics Program [17].

The two groups (breast cancer and benign breast disease cases) were compared by the U-Mann Whitney test [18].

Table 1. Anthropometric characteristic (mean and standard deviation) of pre- and post-menopausal women with breast cancer and benign breast disease (ben breast dis).

	pre-menopausal age				
	n	age yrs	weight kg	height cm	B.M.I. kg/m²
breast cancer	13	43.1	67.8	158	27.4
		8.0	12.3	5.0	4.1
ben breast dis	18	40.1	63.2	157	25.5
		5.8	16.7	6.5	3.7
p-value*		.13§	.08§	.62	.05•
	post-menopausal age				
	n	age yrs	weight kg	height cm	B.M.I. kg/m²
breast cancer	29	61.9	71.6	152	31.2
		8.6	13.2	7.8	5.6
ben breast dis	6	55.6	70.5	151	30.3
		5.6	9.6	6.9	5.8
p-value*		.01•	.98	.51	.61

B.M.I. = body mass index
* by U Mann-Whitney test
• significant different at $p < 0.05$
§ border line significance

RESULTS

In Table 1, the anthropometric characteristic of the women under study are reported. BC women had higher B.M.I. (body mass index) than BBD women in pre-menopausal age (27.4 vs 25.5 kg/m^2; $p < 0.05$), but not in post-menopausal age (31.2 vs 30.3 kg/m^2; p= .61). BC women were older, both in pre- and post-menopausal age.

In Table 2, the serum concentrations of triglycerides, cholesterol, HDL-cholesterol and total lipids are reported. BC women had higher serum concentration of HDL-cholesterol than BBD women both in pre- (52 vs 42 mg/dl; p=0.05) and post-menopausal age (47 vs 40 mg/dl; p = 0.15 border line significance). No difference, instead, between the two groups was observed in serum triglycerides, total cholesterol and total lipids.

DISCUSSION

Several observations collected first in men during cardiovascular disease prevention trials showed an inverse relationship between total serum cholesterol and cancer risk [19].
To elucidate this inverse relationship, it has been hypothesized that cancer, clinically undetectable in its early stage, lowers serum cholesterol [20]. Other authors have suggested that

Table 2. Serum concentrations of triglycerides (TG), cholesterol (CHOL), HDL-cholesterol (HDL) and total lipids (TOTLIP) (mean and standard deviation) of pre- and post-menopausal women with breast cancer and benign breast disease (ben breast dis).

	n	TG mg/dl	CHOL mg/dl	HDL mg/dl	TOTLIP mg/dl
pre-menopausal age					
breast cancer	13	103	215	52	769
		62	160	42	179
ben breast dis	18	111	195	42•	777
		36	34	10	134
p-value*		.60	.58	.05•	.68
post-menopausal age					
breast cancer	29	151	199	47	733
		153	38	10	193
ben breast dis	6	187	202	40	874
		215	27	16	251
p-value*		.86	.83	.15§	.18

* by U Mann-Whitney test
• significant different at $p < 0.05$
§ border line significance

this inverse relationship could be accounted for by higher mortality from cardiovascular disease among those with higher cholesterol level [21]. It has also been suggested that the inverse relationship might be secondary to the strong association between low serum cholesterol and low serum vitamin A [22], which increases the susceptibility to the development of cancer [22].

These hypotheses are of some interest in men but do not explain the direct association between serum cholesterol and breast cancer risk in women reported by some studies [5-7]. A recent prospective study on Norwegian women has reported an inverse relationship between breast cancer risk and serum cholesterol only among cases occurring four years after the collection of blood specimen and not among cases occurring within two years [4]. This observation strongly reduces the likelihood of a "pre-clinical cancer effect" on lipid metabolism in women.

Siiteri [11], and others [12,13] have hypothesized an increased estrogen activity in BC women probably due to increased conversion of cortico-surrenal androgens in estrogens (mainly estrone) by aromatasi of adipose tissue, and to reduced serum concentration of the sex-hormone binding globulin.

The excessive estrogen activity has been involved in the development of breast cancer as promoter of the cancer proliferation and transformation [24].

Estrogens influence serum cholesterol. They decrease low density lipoproteins (LDL)[14], by enhancing the clearance of cholesterol-rich lipoproteins[15]. On the other hand, estrogens increase the serum concentration of high density lipoproteins (HDL) [16].

The effect of estrogens on serum triglycerides is variable because estrogens influence both the production and the removal of triglycerides [15].

In this study, BC women had higher serum HDL-cholesterol than BBD women in pre-menopausal age (52 vs 42 mg/dl; $p=0.05$). Also in post-menopausal age, BC women had higher serum HDL-cholesterol than BBD women (47 vs 40 mg/dl) although the statistical significance of $p=0.05$ was not reached but remained at border line level ($p=0.15$). No difference was observed in the serum concentrations of total cholesterol, triglycerides and total lipids.

The relationship between serum cholesterol and breast cancer could, therefore, reflect estrogen activity. Both associations (inverse or direct) between total serum cholesterol and breast cancer risk are explained by this hypothesis. In fact, the total serum cholesterol in breast cancer patients may be either reduced, due to reduction of LDL-cholesterol, or increased due to the increase of HDL-cholesterol.

A limitation of this study is that the control group is composed of BBD women. Although most benign breast diseases are not pre-malignant, it is likely that women with such abnormalities are at increased risk of developing breast cancer [23]. Therefore BBD women may be not representative of the "normal population". However, as the BBD women are likely to be more similar to breast cancer patients than to the "normal" population, the net effect is probably to underestimate any difference between breast cancer cases and "normal population".

In conclusion, the women with breast cancer had higher serum concentration of HDL-cholesterol than women with benign breast disease, both in pre- and post-menopausal age. This result suggests that the relationship often found between serum cholesterol and breast cancer in women may reflect an increased estrogen activity, rather than a "preclinical effect".

If this hypothesis is further confirmed, HDL-cholesterol could be a biochemical index of increased breast cancer risk in women.

REFERENCES

1. Kark, J.D., Smith A.H., Hames CG., The relationship of serum cholesterol to the incidence of cancer in Evans Country, Georgia. *J Chronic Dis.* : 311-322 (1980).

2. Tornberg, S.A., Holm, L.E., Carstensen, J.M., Breast cancer risk in relation to serum cholesterol, serum ß-lipoprotein, height, weight, and blood pressure. *Acta Oncol.* 27 : 31-37 (1988).
3. Williams, R .R., Sorli, P.D., Feinleib, M., McNamara, P.M. et al., Cancer incidence by levels of cholesterol. *JAMA* 245: 247-252 (1981).
4. Vatten, L.J., Foss, O.P.,Total serum cholesterol and triglycerides and risk of breast cancer: a prospective study of 24239 Norwegian women. *Cancer Res.*50 : 2341-2346 (1990).
5. Wallace, R.B., Rost, C., Burmeister, L.F., Pomrehm, P.R., Cancer incidence in humans: relationship to plasma lipids and relative weight. *J Natl Cancer Inst,* 68 : 915-918 (1982).
6. Dyer, A.R., Stamler, J., Paul, O., Shekelle, R.B. et al., Serum cholesterol and risk of death from cancer and other causes in three Chicago epidemiological studies. *J Chronic Dis.* 34 : 249- 260 (1981).
7. Knekt, P., Reunanen, A., Aromaa, A., Heliovaara, M., Hakulinen, T., Hakama, M., Serum cholesterol and risk of cancer in a cohort of 39000 men and women. *J Clin Epidemiol* 41: 519-530 (1988).
8. Hiatt, R.A., Friedman, G.D., Bawol, R.D., Ury, H.k.,Breast cancer and serum cholesterol. *J Natl Cancer Inst.* 68 : 885-889 (1982).
9. Wingard, D.L., Criqui, M.H., Holdbrook, M.J., Barrett-Connor E., Plasma cholesterol and mortality in an adult community. J Chronic Dis. 37 : 401-406 (1984).
10. Morris, D.L., Borhani, N.O., Fitzimons, E., Hardy, R.J. et al.,Serum cholesterol and cancer in Hypertension Detection and follow up program. Cancer 52 : 1754-1759 (1983).
11. Siiteri, P.K., Hammond, G.L., Nisker, J.A., Increased availability of serum estrogens in breast cancer: a new hypothesis. *In* : Banbury report n 8: Hormones and Breast Cancer : 87-101. Pike MC, Siiteri PK eds, Cold Spring Harbor, New York. (1981).
12. Moore, J.W., Clark, G.M.G., Bulbrook, R.D., Hayward, J.L., Murai, J.T., Hammond, G.L., Siiteri, P.K.,Serum concentrations of total and non-protein-bound oestradiol in patients with breast cancer and in normal controls. *Int J Cancer* 29 : 17-21 (1982).
13. Langley, M.S., Hammond, G.L., Bardsley, A., Sellwood, R.A., Anderson, D.C., Serum steroid binding proteins and the bioavailability of estradiol in relation to breast diseases. *J Nat Cancer Inst* 75 : 823-829 (1985).
14. Kissebah, A.H., Schectman, G., Hormones and lipoprotein metabolism. *in* : Lipoprotein Metabolism. Bailliere's Clinical Endocrinology and Metabolism. Vol 1, No 3, pp 699-727. London; Bailliere (1987).
15. Knopp, R.H., Walden, C.E., Wahl, P.W. et al., Oral contraceptive and postmenopausal estrogen effects on lipoprotein triglyceride and cholesterol in an adult female population: relation to estrogen and progesten potency. *J Clin Endocrinol Metab* 53 : 1123- 1132 (1987).
16. Bradley, D.D., Wingerd, J., Pettiti, D.B. et al., Serum high density lipoprotein cholesterol in women using oral contraceptive, estrogens, and progestins. *New Engl J Med* 299 : 17-20 (1978).
17. National Institutes of Health, National Heart and Lung Institute. Manual of Laboratory Operations. Lipid Research Clinics Program. vol 4. DHEW Publication No. (NIH) 75-628. Bethesda, MD: Government Printing Office (1975).
18. Armitage, P., Berry, G., Statistical Methods in Medical Research. Oxford. Blackwell Scientific Publications (1987).
19. McMichael, A.J., Jensen, O.M., Parkin, D.M., Zaridze, D.G., Dietary and endogenous cholesterol and human cancer. *Epidemiol Review* 6: 192-216 (1984).
20. Rose, G., Shipley, M.J., Plasma lipids and mortality, a source of error. *Lancet* 1 : 523-6 (Lancet).
21. Feinleib M., Review of the epidemiological evidence for a possible relationship between hypocholesterolemia and cancer. *Cancer Res* 43 (suppl): 2503s-2507s (1983).
22. Kark, J.D., Smith, A.H., Hames, C.G., Serum retinol and the inverse relationship between serum cholesterol and cancer. *Br Med J* 284 : 152-154 (1982).
23. Kelsey, J.L., Gammon, M.D., Epidemiology of breast cancer. *Epidemiol Review* 12 : 228-240 (1990).
24. Williams. Textbook of Endocrinology. Wilson JD, Foster DW eds, WB Saunders Company (1985).

CONTRIBUTORS

CONTRIBUTORS

Aaronson, Stuart A.
Laboratory of Cellular and
Molecular Biology
National Cancer Institute, NIH
Bldg. 37 Room 1E24
Bethesda, MD 20892 USA

Ambrosca, Camilla
Istituto di Medicina Interna
e Malattie Dismetaboliche
Clinica Medica
Università di Napoli "Federico II"
Via S. Pansini, 5
80131 Napoli, Italy

Barba, Pasquale
Istituto Internationale di
Genetica e Biofisica, CNR
Via Marconi 10
80125, Napoli, Italy

Beneduce, Gerardo
Istituto per lo Studio e la Cura
dei Tumori
Fondazione "G. Pascale"
Via M. Semmola
80131 Napoli, Italy

Berrino, Franco
Unità di Epidemiologia
Istituto Nazionale per lo Studio
dei Tumori
20100 Milano, Italy

Borrelli, Renato
Nutrizione Clinica, Istituto di Medicina
Interna e Malattie Dismetaboliche
Facoltà di Medicina e Chirurgia
Università di Napoli "Federico II"
Via S. Pansini, 5
80131 Napoli, Italy

Boutron, Marie-Christine
Registre Bourguignon des Cancers Digestives
Equipe Associée INSERM-DGS
Faculté de Médecine
7 Boulevard Jeanne d'Arc
21033 Dijon Cédex, France

Califano Daniela
Istituto per lo Studio e la Cura dei Tumori
Fondazione "G. Pascale"
Via M. Semmola
80131 Napoli, Italy

Catalano, Giuseppe
Oncologia Medica
Dipartimento di Medicina Internistica
Clinica e Sperimentale
II Università di Napoli
Via S. Pansini 5
80131 Napoli, Italy

Celentano, Egidio
Direzione Scientifica
Istituto per lo Studio e la Cura dei Tumori
Fondazione "G. Pascale"
Via M. Semmola
80131 Napoli, Italy

Chiappetta, Gennaro
Istituto per lo Studio e la Cura dei Tumori
Fondazione "G. Pascale"
Via M. Semmola
80131 Napoli, Italy

Ciardullo, Anna V.
Istituto di Medicina Interna
e Malattie Dismetaboliche
Clinica Medica
Università di Napoli "Federico II"
Via S. Pansini, 5
80131 Napoli, Italy

Cillo, Clemente
Istituto Internationale di
Genetica e Biofisica, CNR
Via Marconi 10
80125, Napoli, Italy

Chan, Andrew
Laboratory of Cellular and
Molecular Biology
National Cancer Institute, NIH
Bldg. 37 Room 1E24
Bethesda, MD 20892 USA

Contaldo, Franco
Cattedra di Nutrizione Clinica
Istituto di Medicina Interna e Malattie
Dismetaboliche
Facoltà di Medicina e Chirurgia
Università di Napoli "Federico II"
Via S. Pansini, 5
80131 Napoli, Italy

D'Alessio, Amelia
Centro di Endocrinologia ed Oncologia
Sperimentale del C.N.R.
c/o Dipartimento di Biologia e Patologia
Cellulare e Molecolare
Università di Napoli "Federico II"
Via S. Pansini, 5
80131 Napoli, Italy

D'Amicis, Amleto
Istituto Nazionale della Nutrizione
Via Ardeatina, 546
00178 Roma, Italy

De Filippo, Emilia
Nutrizione Clinica, Istituto di Medicina
Interna e Malattie Dismetaboliche
Facoltà di Medicina e Chirurgia
Università di Napoli "Federico II"
Via S. Pansini, 5
80131 Napoli, Italy

De Vita, Ferdinando
Oncologia Medica
Dipartimento di Medicina Internistica Clinica
e Sperimentale
II Università di Napoli
Via S. Pansini 5
80131 Napoli, Italy

Della Ragione, Fulvio
Istituto di Biochimica delle Macromolecole
Facoltà di Medicina e Chirurgia
Seconda Università di Napoli
Via Costantinopoli, 16
80138, Napoli, Italy

Della Vittoria Scarpati, Michele
Oncologia Medica
Dipartimento di Medicina Internistica
Clinica e Sperimentale
II Università di Napoli
Via S. Pansini 5
80131 Napoli, Italy

Dello Iacovo, Rossano
Direzione Scientifica
Istituto per lo Studio e la Cura dei Tumori
Fondazione "G. Pascale"
Via M. Semmola
80131 Napoli, Italy

de Franciscis, Vittorio
Centro di Endocrinologia ed Oncologia
Sperimentale del C.N.R.
c/o Dipartimento di Biologia e Patologia
Cellulare e Molecolare
Università di Napoli "Federico II"
Via S. Pansini, 5
80131 Napoli, Italy

de Waard, Frits
Department of Epidemiology
Universiteit Utrecht
Postbus 80035
3508 Utrecht, The Netherlands

del Sordo, Gabriella
Nutrizione Clinica, Istituto di Medicina
Interna e Malattie Dismetaboliche
Facoltà di Medicina e Chirurgia
Università di Napoli "Federico II"
Via S. Pansini, 5
80131 Napoli, Italy

Faivre, Jean
Registre Bourguignon des Cancers Digestives
Equipe Associée INSERM-DGS
Faculté de Médecine
7 Boulevard Jeanne d'Arc
21033 Dijon Cédex, France

Farinaro, Eduardo
Cattedra Medicina di Comunità
Facoltà di Medicina
Università di Napoli "Federico II"
Via S. Pansini, 5
80131 Napoli, Italy

Federico, Pasquale
Metodologia Clinica
Dipartimento di Medicina Internistica Clinica e Sperimentale
II Università di Napoli
Via S. Pansini 5
80131 Napoli, Italy

Ferro-Luzzi, Anna
Istituto Nazionale della Nutrizione
Via Ardeatina, 546
00178 Roma, Italy

Fidanza, Flaminio
Istituto di Scienza dell'Alimentazione
Università degli Studi
Via S. Costanzo
06100 Perugia, Italy

Filiberti, Rosangela
Dipartmento di Epidemiologia
e Biostatistica
Istituto Nazionale per la Ricerca
sul Cancro
Viale Benedetto XV, 10
16132 Genova, Italy

Galasso, Rocco
Istituto di Medicina Interna
e Malattie Dismetaboliche
Clinica Medica
Università di Napoli "Federico II"
Via S. Pansini, 5
80131 Napoli, Italy

Ghiselli, Andrea
Istituto Nazionale della Nutrizione
Via Ardeatina, 546
00178 Roma, Italy

Giacosa, Attilio
Nutrizione Clinica
Istituto Nazionale per la Ricerca
sul Cancro
Viale Benedetto XV, 10
16132 Genova, Italy

Guarino, Giuseppina
Metodologia Clinica
Dipartimento di Medicina Internistica
Clinica e Sperimentale
II Università di Napoli
Via S. Pansini 5
80131 Napoli, Italy

Hill, Michael J.
ECP (UK) Headquarters
41 London Street
Andover
Hants SP10 1YN, United Kingdom

Magli, Maria Cristina
Istituto Internationale di
Genetica e Biofisica, CNR
Via Marconi 10
80125, Napoli, Italy

Mazzacca, Gabriele
Cattedra diGastroenterologia e Endoscopia
Facoltà di Medicina e Chirurgia
Università di Napoli "Federico II"
Via S. Pansini, 5
Napoli, Italy

Meyers, Kimberly
Laboratory of Cellular and
Molecular Biology
National Cancer Institute
National Institutes of Health
Bldg. 37 Room 1E24
Bethesda, MD 20892 USA

Miki, Toru
Laboratory of Cellular and
Molecular Biology
National Cancer Institute, NIH
Bldg. 37 Room 1E24
Bethesda, MD 20892 USA

Mineo, Alba
Istituto per lo Studio e la Cura dei Tumori
Fondazione "G. Pascale"
Via M. Semmola
80131 Napoli, Italy

Monaco, Carmen
Istituto per lo Studio e la Cura dei Tumori
Fondazione "G. Pascale"
Via M. Semmola
80131 Napoli, Italy

Oliva, Adriana
Istituto di Biochimica delle Macromolecole
Facoltà di Medicina e Chirurgia
Seconda Università di Napoli
Via Costantinopoli, 16
80138, Napoli, Italy

Palumbo, Rosanna
Istituto di Biochimica delle Macromolecole
Facoltà di Medicina e Chirurgia
Seconda Università di Napoli
Via Costantinopoli, 16
80138, Napoli, Italy

Panico, Salvatore
Istituto di Medicina Interna
e Malattie Dismetaboliche
Clinica Medica
Università di Napoli "Federico II"
Via S. Pansini, 5
80131 Napoli, Italy

Parisi, Valerio
Istituto per lo Studio e la Cura
dei Tumori
Fondazione "G. Pascale"
Via M. Semmola
80131 Napoli, Italy

Perrelli, Andrea
Metodologia Clinica
Dipartimento di Medicina Internistica
Clinica e Sperimentale
II Università di Napoli
Via S. Pansini 5
80131 Napoli, Italy

Quipourt, Valerie
Registre Bourguignon des Cancers
Digestives
Equipe Associée INSERM-DGS
7 Boulevard Jeanne d'Arc
21033 Dijon Cédex, France

Rauscher, Frank J. III
The Wistar Institute of Anatomy
and Biology
3601 Spruce Street
Philadelphia, PA
19104-4268 U.S.A.

Reed, Peter I
Lady Sobell Gastrointestinal Unit
Wexham Park Hospital
Wexham, Slough, Berkshire
SL2 4HL, United Kingdom

Riccardi, Gabriele
Istituto di Medicina Interna
e Malattie Dismetaboliche
Clinica Medica
Università di Napoli "Federico II"
Via S. Pansini, 5
80131 Napoli, Italy

Rossi, Valentina
Metodologia Clinica
Dipartimento di Medicina Internistica
Clinica e Sperimentale
II Università di Napoli
Via S. Pansini 5
80131 Napoli, Italy

Russo, Gian Luigi
Istituto di Biochimica delle Macromolecole
Facoltà di Medicina e Chirurgia
Seconda Università di Napoli
Via Costantinopoli, 16
80138, Napoli, Italy

Santelli Giovanni
Istituto per lo Studio e la Cura dei Tumori
Fondazione "G. Pascale"
Via M. Semmola
80131 Napoli, Italy

Tiberio, Claudia
Istituto Internationale di
Genetica e Biofisica
CNR
Via Marconi 10
80125, Napoli, Italy

Vecchio, Giancarlo
Centro di Endocrinologia ed Oncologia
Sperimentale del C.N.R.
c/o Dipartimento di Biologia e Patologia
Cellulare e Molecolare
Università di Napoli "Federico II"
Via S. Pansini, 5
80131 Napoli, Italy

Visconti, Paola
Nutrizione Clinica
Istituto Nazionale per la Ricerca
sul Cancro
Viale Benedetto XV, 10
16132 Genova, Italy

Zappia, Vincenzo
Istituto di Biochimica delle Macromolecole
Facoltà di Medicina e Chirurgia
Seconda Università di Napoli
Via Costantinopoli, 16
80138, Napoli, Italy

INDEX

INDEX

Printed by Publishers' Graphics LLC USA
DBT140518.23.35.53